D0886780

THE PRINCIPLES OF DAIRY FARMING

Ken Slater

Farming Press

First edition by Kenneth Russell published 1953
Eleventh edition by Ken Slater published 1991
Reprinted 1995

ISBN 0 85236 216 1

A catalogue record for this book is available from the British Library

Published by Farming Press Books and Videos
Wharfedale Road, Ipswich IP1 4LG
United Kingdom

Distributed in North America
by Diamond Farm Enterprises,
Box 537, Alexandria Bay, NY 13607, USA

Cover design by Andrew Thistlethwaite
Phototypeset by Galleon Photosetting, Ipswich
Printed and bound in Great Britain by Butler and Tanner, Frome, Somerset

CONTENTS

PREFACE

When this book was first introduced by Kenneth Russell in 1953, the emphasis was on improving the standard of dairy husbandry. This aspect still continues to be most important, but young people coming into the industry must appreciate that good husbandry must be overlaid by an understanding of the economics of production.

The introduction of quotas in the European Community in 1984 imposed limitations upon production and greatly intensified the necessity for development of skills in dairy herd management. For many years farmers were interested in dairy farming as a way of life, even as a hobby. Everyone can appreciate that hobbies are usually for pleasure and not viewed as a money making exercise.

The current financial difficulties being experienced by many dairy farmers dictate that farming must be thought of as a business. When questioned, many farmers comment that they are 'doing all right'. How can this ever be accepted when the interest charges on overdrafts (which many farmers tend to have) are higher than the rate of return on the capital invested in the dairy business?

The number of dairy farms in the United Kingdom has dropped in the past ten years by 25 per cent, and the number is still going down. The main cause for the decline has been the financial pressures on the industry.

It can never be too early for young persons entering the industry to realise that husbandry processes must be improved continuously in pursuit of better financial rewards. If such a philosophy is taken up then it is guaranteed that dairy farming will become an increasingly exciting life, and the dairy farm business will continue to be financially sound.

I am aware of the increasing complexity of modern dairy production technology; with this in mind I have been fortunate to be able to draw upon the expertise of several friends who are specialists in their areas of dairying. I would like to thank them for their help:

Milk marketing and clean milk production
R Lawton, formerly Head of Dairy Technology Dept., Cheshire College of Agriculture

The genetics of breeding dairy cows
Dr E Wynne Jones, Vice Principal, Harper Adams Agricultural College

Dairy equipment and mechanised feeding
B W Nicholls, Lecturer in Dairy Machines, Cheshire College of Agriculture

Dairy buildings
R. Williams

Veterinary aspects of dairy cows
N M Howie MRCVS

Youngstock rearing
W Thickett BSc, formerly Farm Manager, Barhill Development Farm, BOCM Silcock

With the opening up of the European market in 1992 there are likely to be many exciting challenges developing in the marketing and production of milk. This must result in many avenues for employment and the development of businesses, and the promise of a confident, buoyant milk industry for the future.

Ken Slater
Nantwich, Cheshire
May 1991

CHAPTER 1

THE MILK INDUSTRY

Milk is perhaps the most important single agricultural commodity produced in the UK. From a nutritional point of view it provides the nation with a basic food containing practically all the essential ingredients to promote and maintain life: proteins, carbohydrates, fats, minerals and vitamins. In addition to traditional dairy products like milk, cheese and butter, considerable amounts are also consumed in an ever-widening variety of other products from the food processing, canning, baking, confectionery, dietary and health food industries.

Milk production is economically important in the UK by providing a major sector of food from our own resources. As a single agricultural product, milk accounts for over 20 per cent of the total sales from UK farms. This high proportion of total production is due partly to the UK's temperate climate, heavy rainfall and suitability for growing grass and partly to the systems of subsidy and guaranteed payments.

MARKETING OF MILK

The milk industry as a whole is divided into two main parts: the production of milk, and the processing, manufacture and distributions of milk and milk products. The marketing of milk between these two divisions is organised in the UK nationally by five Milk Marketing Boards (MMB): the English and Welsh MMB, Scottish MMB, the North of Scotland MMB, Aberdeen and District MMB and the Northern Ireland MMB.

To appreciate the present-day situation and discussion on the marketing of milk in the UK, it is helpful to understand some of the background responsible for the formation and development of these boards.

Prior to their formation, dairy farmers individually negotiated the sale of their milk and the price per gallon with their local dairy company. Farmers were also responsible for the transport costs of milk collection and delivery to the purchasing dairy. Farmers therefore paid varying amounts depending on distance to the dairy. Coupled with irregular cash payments and the uncoordinated efforts of the farmers, the whole situation resulted in continued low prices, ineffectual maintenance of incomes and an unstable market.

During the economic depression of the early 1930s, consumer demand for milk and dairy products fell. Dairy companies therefore needed to purchase less milk and terminated supply contracts or negotiated very low prices. Farmers competed by undercutting each other until the milk price fell so low that it became uneconomical to transport resulting in dreadful scenes of wholesome milk being poured down the drain.

The Agricultural Marketing Acts of 1931 and 1933 provided the statutory basis for the formation of agricultural marketing schemes and boards. These were administered by producers and had wide legal powers to market milk and other commodities to the best advantage and in the most economic manner for its producer members.

In 1933, milk schemes were set up in the England and Wales and the Scottish Milk Marketing Board areas, followed in 1934 by the Aberdeen and District, and the North of Scotland schemes. A slightly different scheme was organised in Northern Ireland, but after several modifications, this became subject to the same provisions as the other Boards.

Each marketing Board is controlled by a board comprising mainly of representatives elected by dairy farmers. There are also independent members appointed by the Minister of Agriculture or in the case of Wales, Scotland and Northern Ireland by the relevant Secretary of State. The appointed members were intended to provide the Board with experience and expertise additional to that of elected members. The exact system of electing members varies from Board to Board.

The present system of marketing milk in the UK through regional Boards is based on the following principles:

1. The Boards act as sole buyers of milk from the producers, who have to be contracted to the Board for the sale of their milk under the conditions of the scheme. In return the Boards have a legal obligation to buy and find a market for all the milk offered to them by registered producers, subject to the milk

conforming to necessary standards of quality. Furthermore, the Boards have to market the milk in the most economic manner for producers. Special schemes and contracts apply to producers engaged in 'Farmhouse Cheesemaking', 'Producer-Retailing' and 'On-farm' processing and manufacture.

2. The Boards act as the sole sellers of milk to processors and manufacturers, who must also be contracted to the Board for the purchase of milk under the conditions of the scheme. The Board is, in this sense, a monopoly supplier.
3. The Board is responsible for all transport in collection of milk from farms and in its distribution to processing and manufacturing dairies. This is organised and administered through regional offices using the Board's own transport services or by contracting dairy company transport or private hauliers. All producers pay the same price per litre for transport, irrespective of distance.
4. The Boards organise the allocation and distribution of the national and regional milk supply to processors and manufacturers throughout the country, on the basis of product utilisation priority, the milk price obtainable and the rate of return to the Board. Milk intended for liquid consumption has first priority; it commands the highest price and rate of return to the Board and consequently this market is satisfied first from the available supply. The remaining supply is sold for manufacture of products at different prices related to yield and market returns of the product. Thus cream and specialised soft cheeses would be given higher priority, and command higher prices per litre, than hard cheeses and butter.

   The basic raw material prices paid for different product utilisation are negotiated by a joint committee of the Milk Marketing Board and representatives of the dairy processors and manufacturers. Seasonal adjustments are made as market prices and available milk supplies rise or fall.

   In the past, 60 per cent of the total supply went to the liquid market at the highest price, supporting the lower manufacturing price. Today the situation is reversed, with only 40 per cent going to the liquid market, and new methods of allocation and pricing are currently under review.
5. The Boards collect and pool all the receipts from sales; average the returns per litre; deduct the costs per litre for transport, administration, various capital and/or publicity levies; adjust each producer's price according to milk quality; and pay the

producer once per month by cheque. Today, just as in the 1930s, the monthly milk cheque symbolises stability in the dairy industry.

6. The Boards administer Milk Quality Payment Schemes to improve national milk quality. Each producer's milk is sampled at regular monthly intervals by collection drivers, and tested by the Board's Central Testing Laboratories for compositional and hygienic quality. Producers are paid at a higher than average per litre for good quality milk but are penalised for low quality. The producer's milk quality grade and price differential can have substantial effects on the economics of milk production.
7. Under the 1931 and 1933 Agricultural Acts, Boards were permitted to build processing and manufacturing dairies to utilise surplus milk not taken up by existing dairy companies and to market the products under a brand name. Most Boards retain some processing and manufacturing activity; however in recent years the English and Welsh MMB have separated this function into a separate independent trading company called Dairy Crest.
8. The Boards also offer a wide range of producer services including an advisory service for husbandry and economic milk production problems, milk recording schemes, an artificial insemination service and research into dairy product development.

MILK PRODUCERS—REGISTRATION

Any farmer who wishes to produce milk for sale must first obtain permission from either the Ministry of Agriculture, Fisheries and Food or local authorities, according to where in the United Kingdom he intends to start production. Once authority has been given, the intending producer must register with the appropriate Milk Board and is expected to comply with the regulations concerning clean milk production.

Each Board is required to keep a register of milk producers in its area. This register is available for inspection at the main Board offices. In the England and Wales, Scottish, and Aberdeen areas a milk producer need register only once with the Board even though he may take on the function of wholesale producer, producer-retailer or farmhouse cheesemaker, or any combination of the three. In Northern Ireland and North of Scotland areas a milk producer is required to register in respect of each farm from which he intends to sell milk.

Wholesale Producers

The term 'wholesale producer' is used in England, Wales and the North of Scotland to describe milk producers who sell milk direct to the Board on a wholesale contract. This milk is then disposed of through the commercial outlets by the Boards. The same arrangement applies in the Scottish, Aberdeen, and Northern Ireland areas, where producers are described as 'ordinary producers' or 'milk licence holders'.

Producer-Retailers

'Producer-retailer' is the term used to describe the producer who is authorised by the MMB to retail part or all of his own milk production direct to consumers. Numbers have declined rapidly during the past 35 years, the decline being accelerated by the growth of the large dairy companies distributing heat-treated (pasteurised or sterilised) milk.

Table 1.1 Reduction in Wholesale Producers and Producer-Retailers in England and Wales

Year	*Wholesale Producers*	*Producer-Retailers*
1955	138,305	23,867
1965	99,219	9,961
1975	58,532	4,756
1985	37,432	2,837
1989	32,115	1,910

Source: Dairy Facts & Figures, MMB 1989.

Producer-Processors, Wholesalers, Farm Pasteurisers

This category of producer was introduced by the MMB of England and Wales with an amendment to the Milk Marketing Scheme from 1 April 1981 and was introduced into the Scottish MMB area from April 1982.

The arrangement is for a producer-processor to be allowed to sell his own heat-treated milk in a retail container to someone other than the ultimate consumer, that is, a shop, supermarket or dairyman.

Any milk withheld from the Board must be under the terms of a specific agreement with the Board and the producer must make payment to the Board for the amount of milk involved.

This payment is an amount which leaves the producer with the equivalent of the average wholesale producer price plus amounts equal to the average ex-farm haulage cost and bulk addition, and without any deduction for co-responsibility levy. As a processor he obtains the normal margins.

There is no Board requirement for sampling and testing the supply for compositional quality payment purposes or for the antibiotics deduction scheme.

Farmhouse Cheesemakers

Farmhouse cheesemakers are milk producers who make cheese on the farm under a special contract with the Board. The Board is responsible for grading Cheddar, Cheshire and Lancashire 'farmhouse' cheeses and for selling them, on behalf of the makers, to recognised cheese factors. The number of farmhouse cheesemakers in England and Wales dropped from a pre-war total of over 1,000 to 193 by 1984.

THE EFFECT ON THE DAIRY INDUSTRY OF ENTRY INTO THE EUROPEAN COMMUNITY (EC)

When the United Kingdom moved towards entry into the EC, it seemed likely that existing legislation would threaten the future of the Milk Marketing Board. Eventually after much discussion, Community regulations were amended to ensure that the essential powers of the Boards could be maintained provided certain conditions were met. A referendum of producers in 1978 gave an overwhelming vote in support of the continuation of the Boards. The Boards have continued to work to the present period, but their long-term future comes up for further discussion in the early 1990s.

The biggest impact on British dairying as a result of entry into the EC was the quota system.

The Quota System

This system was introduced in 1984 as an attempt to correct the imbalance between milk output and consumption in the EC. A quota for milk production within the Community was originally introduced for a period of five years; this was extended by agreement for a further three years until 1991–92. With a few minor exceptions the quota system applies to all sales of milk or milk products from farms. Quota is divided into two categories:

1. Direct sales quota—covering the sale of milk and milk products by the farmer directly to the consumer.
2. Wholesale dairy quota—covering deliveries of milk from the farm to the dairy.

The EC allocated a wholesale and direct sales quota to each member state (see Table 1.2). This was then distributed to the nation's farmers. In the UK a direct sales quota is allocated to individual producers whereas the wholesale quota is divided up by the government, and allocated to each of the nine geographical regions. Small amounts of quota have been left unallocated in order to form a reserve.

Producers are made to conform to the quota by means of a levy system.

The Levy System

Initially levy was payable on all milk produced in excess of quota. Subsequently the rules have been altered so that only those producers most over quota are liable for levy. This is discussed later in this chapter.

The rate of levy is set as a percentage of the Target Price. The target price is a delivered-to-dairy price for milk of 3.7 per cent butterfat. It is agreed each year by the EC council of agricultural ministers, and was 19.64 pence per litre (ppl) in 1988–89.

Levy payments by purchasers The purchasers in the UK are the 5 Milk Marketing Boards and are responsible for all levy payments to the EC. Assessment for the levy is initially a comparison between the total amount of milk delivered and the quota assigned to each Board. Any excess is subject to a levy payment. This money is then recoverable from those producers who are themselves over quota. Since quota regulations were initially

Table 1.2 Distribution of Quota to EC Member States 1988–9

EC Member State	Wholesale Quota							
	Guaranteed Quantity for 1988–9	Guaranteed Quantity for 1986–7	Temporary Suspension (5.5% of B)	Actual Reference Quantity	Direct Sales Quota	Total: Wholesale plus Direct	Reserve Quota	Total Quota (including reserve)
	A	B	C	D=A−C	E	F=D+E	G	H=F+G
	thousand tonnes							
Belgium	3,121.861	3,211	176.605	2,945.256	380.809	3,326.065		3,326.065
Denmark	4,735.540	4,882	268.510	4,467.030	0.970	4,468.000		4,468.000
France	24,964.980	25,634	1,409.870	23,555.110	747.780	24,302.890		24,302.890
Germany	22,753.310	23,423	1,288.265	21,465.045	93.100	21,558.145		21,558.145
Greece	520.890	537	29.535	491.355	44.620	535.975		535.975
Ireland	5,121.600	5,280	290.400	4,831.200	15.520	4,846.720	303.000	5,149.720
Italy	8,534.060	8,798	483.890	8,050.170	1,082.520	9,132.690		9,132.690
Luxembourg	257.050	265	14.575	242.475	0.970	243.445	25.000	268.445
Netherlands	11,619.630	11,979	658.845	10,960.785	92.150	11,052.935		11,052.935
Spain	4,560.500	4,650	255.750	4,304.750	677.500	4,982.250	50.000	5,032.250
UK	14,869.687	15,329.574	843.127	14,026.560	383.563	14,410.123	65.000	14,475.120
Total	101,059.108		5,719.372	95,339.736	3,519.502	98,859.238	443.000	99,302.235

Source: Dairy Facts & Figures, MMB 1989.

introduced, some amendments have been made—re-allocation of unused quota is now allowed between producers, purchasers and regions. The re-allocations cannot take place until after the end of the quota year in March. The rate at which the levy is to be paid is announced annually in July. In 1989 the levy on wholesale milk in excess of quota was 100 per cent of the target price.

Levy payments on direct sales A producer who makes direct sales in excess of his direct sales quota is also liable for a levy payment. If after re-allocation and interchange of quota it is shown that UK direct sales are less than the total national quota no levy is paid. In 1989 levy on direct sales milk was assessed at 75 per cent of the target price, i.e. 14.73 pence per litre.

Levy payment on wholesale sales Those producers whose sales remain within their quota can never be subject to levy payments. Where a purchaser (i.e. the Milk Marketing Board) has not had to pay levy then none of the producers delivering to that purchaser can be asked for levy even if they have individually exceeded their quota. Where a purchaser pays levy then it will have to recover the payment from its wholesale producers.

The system for levy payments was changed after 1988, by the introduction of the *Threshold*. Under the current system only the producers who exceed their quota by more than the threshold percentage pay levy. The percentage beyond which the producer starts to pay the levy is calculated according to the total production which exceeds the threshold. This provides the money to reimburse the purchasers for the levy paid to the EC.

The degree of over production in each Board area will result in different threshold levels being applied to the wholesale producers in each area. As the production year draws to a close so each Board produces provisional threshold levels as a guide for milk producers, e.g. a threshold of 7.5 per cent may be suggested as a warning not to go beyond that level and to those who are already over producing to cut down on production. The final threshold level is not confirmed until some two months after the end of the quota year. The last months of the quota year are often a very exciting, even terrifying time for many producers. While some farmers cut back on over-threshold production others increase theirs hoping that the threshold will actually rise.

An example of a typical over-quota situation could be described as follows:

A dairy farm with a quota of 500,000 litres plus a confirmed threshold after the end of the quota year at 7.5 per cent (i.e. 37,500 litres) would be allowed to produce 537,500 litres of milk.

Where the actual production of milk was 560,000 litres, this would result in 22,500 litres of milk in excess of the quota + threshold amount.

In 1988–89, the penalty was 19.64 pence per litre.

Total levy payable would be £4,419.

Butterfat—Effect on Quotas

In 1986, an additional feature to the quota situation was introduced—milk volume was linked to butterfat content. Higher than normal butterfat levels in milk receive higher payment, and thus many farmers pursued higher levels of butterfat to increase milk income which had been reduced by the restrictions on milk volume caused by the quota system.

To prevent this development, each producer was given a butterfat reference figure associated with his volume level of quota milk. The producer's production is adjusted up or down for quota purposes by the amount that his butterfat average differs from his reference level. The butterfat reference figure is referred to on the monthly milk statement as the farmer's annual butterfat base. The reference figure corresponds to the average level of butterfat that was associated with the herd during the year prior to the introduction of quotas.

An 0.18 per cent adjustment to milk volume is made for every 0.01 percentage point difference in butterfat.

It is the adjusted milk volume that is used to judge whether a producer has exceeded his milk quota. This development means that the butterfat content of a producer's milk must be recorded in addition to the volume of the milk supplied by the producer. The monitoring of the butterfat level has to be carried out by the purchaser (i.e. the dairy).

Adjustment to a producer's figure is only triggered when the purchaser's butterfat level is exceeded. The purchaser's own butterfat reference level is the average butterfat level of milk purchased in the year prior to quotas.

Triggering happens when the average butterfat level of milk delivered to the purchaser exceeds the purchaser's own butterfat reference level. If the average is equal to or less than the purchaser's reference level, no downward adjustment is made for butterfat and there will be no adjustment to individual producer's delivery figures.

A further revision of some of the rules was introduced in 1989–90. From this date, the triggering of butterfat adjustments will only happen if the average butterfat level for all UK deliveries exceeds the overall UK reference level. A producer will only have a potential adjustment triggered if both the UK as a whole and his specific Board exceed their respective reference levels.

Leasing of Quotas

In recent years, more and more advantage has been taken of the quota leasing scheme. There were 10,000 applications made for leasing in 1989/90 and the total for 1990/91 is likely to exceed 13,000. Any leasing arrangements must normally be completed by the end of July each year. Leasing offers an opportunity to make a quota fit a farm, rather than having the farm correspond to a quota which in many cases could be financially crippling. Leasing applies to the current quota year only, with the quota reverting to its original owner at the commencement of the following year. Leasing costs in the range of 5 to 8p per litre. The system of leasing has given a considerable degree of flexibility to the quota system. A keen young farmer would be able to lease quota, and use the extra income generated to possibly buy additional quota thereby increasing the long term production potential of his business.

Permanent Transfer of Quota

The sale of quota is a very necessary part of the quota system. The transfer arrangements are rather complex and professional advice should be sought. There are several instances when quota transfer is likely, e.g. the sale of all or part of a dairy farm, the inheritance of a farm, and the formation or dissolution of partnerships.

Transfers can be effected at any time during the quota year. At the commencement of a quota year, any quota bought in would be described as 'clean'. This means that it would be 100 per cent available to the purchaser for that quota year. As the year progresses, so the percentage available tends to fall to nil by the year end. The cost of buying quota will be set according to its percentage use.

There are very large fluctuations in the demand for quota from year to year. Favourable climatic conditions provide good forage which leads to increased milk production, probably well beyond quota levels. It is in this situation that prices for quota can rise to very high levels. The range of sale price has been as wide as 30–50p in recent years.

When quotas were first introduced, confidence in the future of the industry in the UK dropped to a very low level. This resulted in the price of land suited to dairy farming dropping from £7,500 per hectare to £3,750 per hectare. Once the quota regulations had been assimilated and understood by the industry, it was realised that the potential sale value of quota, at 30p per litre on a yield of 12,500 litres per hectare, represented a total value of £3,750—replacing the value of the capital asset lost by the fall in land values. The EC quota system is to be reviewed in 1992 and the future of the quota system is to be decided.

Co-responsibility Levy

In 1977 the EC established a co-responsibility levy on milk to create a fund to be used for schemes designed to expand the market for milk and milk products produced in the Community. This levy is set each year as a percentage of target price and is, therefore, subject to variation. It is paid by producers on all milk delivered to dairies; direct sales are exempt. Milk producers who farm in less-favoured areas (LFA) are exempt from paying the levy. Producers elsewhere who delivered less than 60,000 litres in 1987–88 or who had a quota of less than this amount on 1 April 1989 pay a reduced rate.

The co-responsibility levy from February 1990 was 0.3053 pence per litre.

ORGANISATION AND TRENDS IN DAIRY PRODUCTION

The structure of dairy farming has altered considerably during the past thirty years.

1. From 1973, the Dairy Herd Conversion Scheme followed by the Non-Marketing of Milk Scheme, had a significant influence on the decline of milk producers. Between 1975 and 1989 the number of producers declined by 45 per cent. (See Table 1.1.)

2. Although the number of producers has declined greatly over the years, the number of dairy cows in the country has decreased by only 14 per cent during the same period. (See Table 1.3.)

Table 1.3 Numbers of Dairy Cows in the UK

	Dairy Cows			
June Census	*England & Wales*	*Scotland*	*Northern Ireland*	*United Kingdom*
		(thousands)		
1965	2,650	340	196	3,186
1970	2,714	321	208	3,244
1975	2,701	302	239	3,242
1979	2,734	290	264	3,288
1980	2,672	282	270	3,224
1981	2,638	278	271	3,187
1982	2,684	282	280	3,246
1983	2,745	289	294	3,328
1984	2,696	283	299	3,278
1985	2,580	273	294	3,147
1986	2,573	270	292	3,135
1987	2,486	264	289	3,039
1988	2,379	251	279	2,908
1989	2,341	247	280	2,868

Source: Dairy Facts & Figures, MMB 1989.

3. Cow numbers in general increased throughout the UK during the period 1965–83 (see Table 1.3). This trend, coupled with reducing numbers of producers, inevitably resulted in larger herd sizes. (See Table 1.4.)

   When quotas were introduced in 1984, cow numbers decreased, particularly in England and Wales, and they have gone down fairly steadily ever since.
4. Average milk yield has improved considerably over the years (see Table 1.5). It would be very difficult to pinpoint the reasons for this increase but in broad terms it must be attributed to greatly improved management techniques, which in turn were able to exploit and realise the greatly increased genetic potential of the national dairy herd for milk production.

Table 1.4 Average Herd Size in England and Wales

Size Group (number of dairy cows)	*England and Wales* 1982	1987	1988
	Per cent of total herds		
3–9	4.7	3.8	4.0
10–19	8.7	7.2	7.2
20–29	11.7	9.8	9.8
30–39	12.2	11.7	12.0
40–49	11.7	12.2	11.9
50–59	9.8	10.3	10.2
60–69	8.6	8.8	8.7
70–99	16.7	18.1	18.1
100–199	13.6	15.7	15.6
200 and over	2.3	2.4	2.4
Total	100.0	100.0	100.0
Total herds ('000)	43.0	37.4	35.9

Source: Dairy Facts & Figures, MMB 1989.

Table 1.5 Average Annual Yield per Cow in the UK

MILK YIELD PER COW

April to March	*England and Wales*	*Scotland*	*Northern Ireland*	*United Kingdom*
	litres			
1969–70	3,755	3,750	3,660	3,750
1974–5	4,070	4,045	3,795	4,050
1978–9	4,680	4,570	4,415	4,650
1979–80	4,715	4,535	4,325	4,670
1980–1	4,810	4,600	4,410	4,760
1981–2	4,800	4,765	4,450	4,745
1982–3	5,085	4,990	4,830	5,055
1983–4	4,950	4,970	4,850	4,940
1984–5	4,765	4,860	4,575	4,770
1985–6	4,930	4,865	4,600	4,880
1986–7	4,985	5,005	4,550	4,945
1987–8	4,870	4,990	4,635	4,870
1988–9	4,915	5,015	4,585	4,895

Source: Dairy Facts & Figures, MMB 1989.

The use of artificial insemination has brought the availability of good-quality, proven bulls within the scope of every milk producer in the country and has thus enabled the production potential of the national dairy herd to be raised so quickly.

Table 1.6 Number of Inseminations by Breed from MMB Centres in England and Wales

	England and Wales	
	1987–8	*1988–9*
Breed of Bull	*thousands*	
Dairy		
Ayrshire	13	13
Dairy Shorthorn	5	5
Friesian/Holstein	1,040	1,047
Guernsey	12	12
Jersey	22	21
All Dairy	**1,093**	**1,099**
Beef		
Aberdeen Angus	62	68
Beef Shorthorn	—	—
Belgian Blue	116	112
Blonde d'Aquitaine	29	31
Charolais	217	222
Chianina	—	—
Devon	4	4
Hereford	161	143
Limousin	337	281
Murray Grey	7	7
Romagnola	—	—
Simmental	86	98
South Devon	4	4
Sussex	4	4
Welsh Black	7	8
Others	1	1
All Beef	**1,035**	**984**
Other Breeds	**3**	**7**
All Breeds	**2,131**	**2,091**

Source: Dairy Facts & Figures, MMB 1989.

Table 1.7 Milking Systems – England and Wales

	England and Wales		
Type of System	*1981*	*1984*	*1988–89*
Cowshed			
Hand	228	164	75
Bucket	6,229	4,426	2,377
Pipeline	8,563	7,403	5,672
Total Cowshed	**15,020**	**11,993**	**8,124**
Parlour			
Herringbone	14,273	15,651	15,170
Abreast	11,831	11,849	9,880
Tandem / Chute	506	416	293
Movable Bail	254	236	164
Rotary	289	223	126
Others	—	—	—
Total Parlour	**27,153**	**28,375**	**25,633**
Total	**42,173**	**40,368**	**33,757**

Source: Dairy Facts & Figures, MMB 1989.

5. Since 1955 the breed pattern of the national dairy herd has changed, the proportion of Friesian/Holsteins increasing from 34.5 per cent in 1971 to 95 per cent of dairy breeds in 1989, and is progressing further along this path. Some indication of the current use of breeds in milk production can best be provided by the number of inseminations by breeds at the insemination centres.

   Many of the traditional breeds have lost much ground to newly imported breeds, particularly in the beef section. The increasing production of beef as a by-product from the national dairy herd has accelerated the trend towards the introduction of the Continental breeds. The shift in the pattern of cattle breeds in general has certainly been due to the economic pressures felt by the dairy industry over the years since the Second World War. (See Table 1.6.)
6. A development arising directly from the increases in herd size and yield, coupled with an increasing cost of labour, has been the changes in milking systems to be found on many farms.

The trend has been to more sophisticated methods, so that by 1989, 76 per cent of the herds in England and Wales were milked through some version of a milking parlour. (See Table 1.7.) During recent years there has been a distinct movement towards automation in the milking systems, in an effort to eliminate routine mechanical tasks and so leave the herdsman more time to concentrate on husbandry.

MILK COLLECTION AND SAMPLING

All ex-farm milk supplies are now bulk collected in England and Wales. In the days of churn collection, a milk lorry driver would check that each churn contained milk of a satisfactory nature. Now with the bulk tank system in operation, a greatly increased responsibility rests on the tanker drivers to ensure that (i) sub-standard milk which might contaminate a whole tanker load is not bulked with other supplies; (ii) a representative sample is obtained for testing; and (iii) each producer's supplies are correctly recorded.

The main features which are looked for in the initial inspection of the bulk tank supply before collection are:

(a) The milk must not have bad or sour smells, and no blood or foreign matter on the surface.
(b) There must be no smells, foreign matter or blood evident after the milk has been agitated in the tank for at least two minutes.
(c) The milk temperature should be within the Board's guidelines (these vary between Boards).

If a consignment does not meet with these requirements, then the tanker driver can consult his depot and this can lead to the milk being rejected.

MILK QUALITY SCHEMES

Central Testing of Milk

The quality of milk in all its different aspects has increasingly become an important part of the selling of liquid milk for human consumption since the setting up of the Milk Marketing Board in October 1933.

Table 1.8 Hygienic Quality TBC – Incentive/Penalties. England and Wales MMB

Band	*Criteria* a	*Incentive pence per litre* c
A	Where the average TBC is 20,000 or less	+0.230
B	Where the average TBC is above 20,000 but not more than 100,000	nil
C 1	Where the average TBC is above 100,000 and there has been no deduction in the previous six months	−1.500 *b*
C 2	Where the average TBC is above 100,000 and a C1 deduction has applied in the previous six months	−6.000 *b*
C 3	Where the average TBC is above 100,000 and a C2 or C3 deducation has applied in the previous six months.	−10.000 *b*

a Average TBC is based on two or more valid tests for the month. If there are less than two such tests, the supply will be placed in Band A if this was the band applicable in the last month in which the producer consigned milk, otherwise it will be placed in Band B.
b Provided at least two test results are above 100,000; otherwise the supply will qualify for Band B.
c From 1 April 1989.
Source: Dairy Facts & Figures, MMB 1989.

The two areas for development in milk quality are hygiene and compositional quality.

The task of monitoring the quality of the National milk supply provided a tremendous problem of organisation and administration resulting in the setting up in 1982 of six Central Testing Laboratories to be run by the Milk Marketing Board, instead of the previous 250 or so dairy and creamery laboratories. The Central Testing Laboratories are situated at Harrogate, Newcastle-under-Lyme, Llanelli, Thames Ditton, Yeovil and Plymouth.

Hygienic Quality Schemes

The joint committee of the MMB and the Dairy Trade Federation agreed Quality Control conditions which authorise the testing of the marketability of any producer's supply sold under wholesale contract. Initial inspection of bulk collected supplies is by the tanker driver, and if there is doubt about quality, a sample may be taken and tested before collection.

Once the Central Testing of ex-farm milk was introduced over the Board area, it was possible to move into a hygiene scheme based on Total Bacterial Count (TBC). This replaced the old scheme based on the Resazurin Test.

Each producer's milk supplies are tested once per week and the average of the TBC results in a month is used to place the supply into one of several hygiene quality bands, with each band receiving a different price. Any producer who persistently supplies milk of poor quality is reported to the Board and may have his milk supply contract cancelled.

The introduction of TBC testing has greatly improved the hygienic quality of ex-farm milk; it has also had a marked effect in reducing sediment levels in milk. Since April 1983 the sediment testing scheme has reverted to a voluntary basis.

Table 1.9 Hygienic Quality TBC – Incentive/Penalties. Scottish MMB

Criteria a	*Incentive* *pence per litre* b
Where the average TBC is below 15,000 and where not more than two available results are 15,000 or above	+0.15
Where the average TBC exceeds 50,000 and at least two individual non-consecutive results exceed this level (but the conditions for the higher penalty rate below are not met)	−2.0
Where the geometric average TBC of tests over the preceding two months exceeds 100,000 and at least two test results exist for each of those months	5.0
All other cases	nil

a Average TBC is based on three or more valid tests for the month. If only two results are available, a third result of zero is assumed, thus giving the producer the best possible situation. If only one test result is recorded, the incentive is nil.
b From 1 April 1989.
Source: Dairy Facts & Figures, MMB 1989.

Table 1.10 Hygienic Quality TBC – Incentive/Penalties. Aberdeen & District MMB

Criteria a	*Incentive pence per litre* b
Where the average TBC does not exceed 45,000	nil
Average TBC between 46,000 and 90,000	−0.243
Average TBC between 91,000 and 130,000	−1.215
Average TBC exceeds 130,000	−2.430

a Average of three or more valid tests for the month.
b From 1 April 1984.
Source: Dairy Facts & Figures, MMB 1989.

Table 1.11 Hygienic Quality TBC – Incentive/Penalites. North of Scotland MMB

Band	*Criteria* a	*Incentive pence per litre*
A	Where the average TBC is not greater than 50,000	nil
B	Where the average TBC exceeds 50,000 and at least two individual non-consecutive results exceed this level but the conditions for a higher rate penalty are not met	−0.5
C	Where the average TBC exceeds 100,000 and at least two individual non-consecutive results exceed this level but the conditions for a higher rate penalty are not met	−1.0
D	Where the geometric average TBC of tests over the preceeding two months exceeds 100,000.	−10.0

a Average of three or more valid tests for the month. If only two results are available, it is assumed that a third result of zero is available, thus giving the producer the best possible situation. If only one test result is recorded, the incentive is nil.
Source: Dairy Facts & Figures, MMB 1989.

Table 1.12 Hygienic Quality TBC – Incentive/Penalties. Northern Ireland MMB

Class	*Criteria* a		*Incentive pence per litre*
H+2	less than	26,000	+0.12
H+1	26,000 –	50,000	+0.06
H	51,000 –	100,000	nil
H−1	101,000 –	300,000	−1.2 *b c*
H−2	more than	300,000	−3.0 *b c d*

a Average TBC: in computing the average, each result is placed in the appropriate payment class and if the highest result for a supply is the only one in that class, it is discarded and the remaining results averaged. If there is only one result, the supply is placed in that class most appropriate given previous months' results.
b These penalties are reduced to 0.6 ppl if the supply has not incurred any penalty in the preceding eleven months.
c An additional penalty of 3.0 ppl will apply if the supply was over 100,000 in each of the preceding two months or in any five of the preceding eleven months.
d The additional penalty referred to in footnote *c* will be 6.0 ppl if the supply was over 300,000 in each of the preceding two months or in any five of the preceding eleven months.
Source: Dairy Facts & Figures, MMB 1989.

ANTIBIOTICS IN MILK

In recent years there has been a considerable growth in the use of antibiotics for treatment of cows suffering from bacterial infections. Mastitis has been established as a very important cause of reduced milk production, and its treatment has become closely related to the use of antibiotics. Intramammary treatment is the routine method of mastitis control.

Another area of antibiotic use relating to milk production is 'dry cow therapy', that is, antibiotics administered at the drying-off stage. Whilst a cow is in milk, any administered antibiotics can leave traces in the milk for up to four days after treatment.

Significant amounts of antibiotic residues may be present in the milk immediately after calving and for longer than four days if the manufacturer's advice is not followed. The testing for antibiotics in milk has become increasingly important because:

(i) A small percentage of the human population are very sensitive to the effects of antibiotics, particularly penicillin.

(ii) The presence of antibiotics is very harmful to the bacteria in milk. This is particularly important in milk which has to be subjected a fermentation process, e.g. cheese or yogurt. The MMB requires producers to withhold from sale the milk from treated cows until the recommended period has elapsed.

Details of prices paid in 1989 for consignments failing an antibiotics test are given in Table 1.13.

Table 1.13 Prices paid for consignments failing an antibiotics test

England and Wales	**1.0 ppl**
Scottish MMB	**12.00 ppl** for a first failure **3.00 ppl** for subsequent failures A first failure is where there has been no failure in the previous six months.
Aberdeen and District	**5.00 ppl** for the first or second failure within the past 12 months. **3.00 ppl** for a third or fourth failure within such a period. **1.00 ppl** for a fifth or subsquent failure within such a period. Tests are carried out daily until a negative result is obtained.
North of Scotland	**2.00 ppl** Samples are tested daily until two consecutive clear results are obtained.
Northern Ireland	**75 per cent** of the month's price for a first failure within the past 12 months. **60 per cent** of the month's price for a second failure within such a period. **10 per cent** of the month's price for a fourth or subsequent failure in such a period. *The following also applies:* **Suspension of the producer's Wholesale Contract for 2 days** for a fifth failure within the 12 month period. **Suspension for 4 days** for a sixth failure in such a period. **Suspension for 6 days** for a seventh or subsequent failure in such a period.

Source: Dairy Facts & Figures, MMB 1989.

SAMPLING FOR DISEASES

Brucellosis

The condition in cattle is a contagious disease which can cause abortion, infertility and lowered milk yield due to failure to breed and an extended calving index. Once identified infected animals are culled. In October 1985 all herds in the UK were designated officially free of brucellosis. All herds are periodically tested by the various Departments of Agriculture to ensure that they remain free from infection.

Dairy herds have milk samples tested monthly by the Milk Boards on behalf of the Departments. If any herds fail the test, there is a further investigation by the Departments. All abortions or premature calvings must be reported and animals isolated pending tests.

Mastitis Cell Count

The sample of milk used for brucellosis testing is also used for monitoring the herd mastitis cell count. The result is entered on to the producer's payment advice together with the rolling 12-month average. Where there are problems due to mastitis, the MMB provides a control service.

PRIMARY SCHEMES FOR MILK PAYMENT

The proportion of milk going into the manufacturing section of the industry is steadily rising. The importance of this factor was reflected in changes introduced by the Milk Marketing Board of England and Wales on 1 April 1984. Three major changes were made:

1. Compositional Quality

Milk is paid for on the basis of compositional quality, assessed according to the various levels of fat, protein and lactose. The values placed on each constituent are related to their relative market values and may be revised at any time. They are also subject to retrospective increases during the year which the producers receive in the form of supplementary payments for milk previously delivered. The level of payment for constituent values of milk in 1990 (summer) is shown in Table 1.15.

2. Contemporary Payments

Payments for compositional quality are made on test results obtained within the month instead of using the average of previous months—known as the historical average.

This is an attempt to relate milk payment to the current quality level and to provide an incentive for producers to achieve better quality results for a more immediate price improvement.

3. Seasonal Adjustments

Seasonal price differentials have been introduced to promote a more even supply of milk. There is a marked decline in milk production during late summer/early autumn which in conjunction with the increased milk production of spring/early summer causes problems of over and under supply.

At times of peak production an increased processing capacity is required which cannot be utilised at other times of the year. Orderly marketing of dairy products is also difficult when the supply situation is so variable.

To change the seasonality of production at a slightly faster pace, alterations were made from April 1989 to move away from

Table 1.14 Seasonal Price Differentials in Wholesale Milk, England and Wales

Month	*From 1 April 1990 per cent*
April	− 2.9
May	−14.5
June	−11.6
July	+21.6
August	+30.3
September	+25.7
October	+21.3
November	+ 2.6
December	+ 2.6
January	+ 2.6
February	+ 2.6
March	+ 2.6

Source: Dairy Facts & Figures, MMB 1989.

a pence-per-litre scheme and towards a percentage adjustment to the producer's price (see Table 1.15).

The payment systems followed in each Board area in the UK may have slight variations in detail but basically follow the same formulae as shown in the example. Full details can always be obtained from the appropriate Board offices.

Table 1.15 Calculation of Milk Price, England and Wales (July 1990)

Assumed composition of milk			*Assumed constituent values*	*Producer's price*
	%		*ppl per 1%*	*ppl*
Butterfat	3.95	×	2.141	8.456
Protein	3.21	×	2.254	7.235
Lactose	4.55	×	0.329	1.496
			Total	**17.187**
			Allowance for seasonality (21.6%)	+3.712
				20.899
			Adjusted for hygiene (TBC) Band A	+0.230
			Total	**21.129**
			Adjusted for co-responsibility	−0.305
				20.824

CHAPTER 2

REGULATIONS GOVERNING MILK PRODUCTION

From the point of view of the consumer, processor and manufacturer, milk quality is signified by the following characteristics which eventually determine its suitability for human consumption:

- Good composition in terms of butter fat, solids-not-fat and lack of adulteration by added water.
- Freedom from abnormal milks, colostrum, mastitic milk and from microbial, food, weed or absorbed taints.
- Freedom from antibiotics, residues of detergent and chemical sterilants and other unnatural substances.
- Freedom from physical dirt, sediment, and, as far as possible, blood, body cells and mucus.
- Freedom from pathogenic microorganisms affecting cattle and man.
- Freedom from as many other microorganisms as possible which could give rise to faults, poor keeping quality and low marketability.

Milk production legistration is mainly concerned with milk hygiene and the protection of public health. Other quality aspects, rules and standards are also required to be met under the conditions of sale contract with the relevant Milk Marketing Boards.

Before a producer can sell milk in the UK he must first be registered with the Ministry of Agriculture. The conditions of production on the farm must also comply with the provisions of the latest Milk and Dairies (General Regulations) order. Furthermore the producer is required to be contracted to his relevant MMB under the Marketing Acts and is bound by the sale conditions and its rules.

THE RULES SUMMARISED

Registration

All dairy farmers and the premises which they use must be registered. The county dairy husbandry officer (of the Ministry of Agriculture) will inspect the farm on behalf of the licensing authority and see that the milk is being produced in a satisfactory manner. MAFF can refuse or cancel registration if it appears that the regulations cannot be or are not being complied with. In addition, contravention of the regulations is an offence for which there may be legal penalties.

Buildings

The regulations regarding dairy premises lay down quite strict provisions regarding the siting and construction of buildings, as not to give rise to contamination of milk. The air in dairy premises must be kept in a fresh and wholesome condition by adequate ventilation. There should be adequate illumination for all dairy operations to be carried out in good light.

Floors and walls liable to soiling must be impervious and constructed in such a way that they can be easily cleaned. Floors should be sloped and provided with channels so that liquids can be conveyed to a suitable trap drain outside the building and thence to a suitable place of disposal.

Milking houses and milk rooms must be kept clean and the access and surroundings to these buildings must be free of any accumulation of dung or other offensive matter. Direct access to or by pigs, poultry or calves should also be prevented.

Milking bails are able to conform to these regulations if moved sufficiently frequently to avoid contamination or if sited on a concrete platform.

Water Supply

The regulations require that there must be a suitable and sufficient water supply on the dairy farm and adequate precautions taken against the risks of its becoming polluted or contaminated.

Production Methods

Provisions are laid down to act as precautions against unnecessary contamination of milk. Movement of dry bedding, hay or other dusty matter in the milking parlour should be avoided within half an hour before milking begins, or during milking. Cows' flanks, tail, udder and teats should be kept clean by washing, clipping and grooming. The udder and teats should be washed and cleaned prior to milking, and kept clean during milking.

The foremilk of each cow must be drawn into a separate receptacle for visual examination and subsequently discarded in such a manner as to avoid risk of infection.

The hands of the milker must be thoroughly washed and dried before milking begins and should be kept clean, dry and free from contamination during milking. Milkers must wear clean and washable overalls, wear head coverings and keep suitably covered any cut or abrasion; they should also refrain from smoking or spitting during milk handling.

No vessel or appliance may be used for milk if it cannot be easily cleaned. Every vessel and appliance brought into contact with milk must be thoroughly clean prior to use. All vessels containing milk must have a suitable cover.

Following milking the milk must be strained and cooled without delay. MMB sale conditions stipulate that it should be cooled to less than 7°C in refrigerated bulk tanks and held at this temperature until collection. Sale conditions also require that producers refrain from adding the following to the bulk milk: water, mastitic milk, milk produced within 72 hours of calving or milk from cows which have been treated with antibiotics in the previous 72 hours. Failure to comply with these conditions may lead to non-acceptance of the bulk milk or financial penalisation, following test failures on the Board's milk quality payment schemes.

If a person engaged in dairy work becomes aware that he, or a member of his household, or the household of any of his employees, is suffering from a notifiable human disease—salmonella, typhoid, tuberculosis, scarlet fever or dysentery—he must notify the local Medical Officer of Health. If the medical officer is not satisfied, or has good grounds for suspecting that a supply of milk is infected, he may stop the sale of milk from the farm or require it to be treated in such a way that it will be rendered safe.

Powers are given to officers of the Ministry of Agriculture to visit farms, inspect cattle, take samples and inspect premises and methods to see that they comply with the regulations.

METHODS THAT MATTER

Clean milk production is based on two main aspects:

(i) Keeping microorganisms, physical dirt and abnormal milk out of the bulk milk.
(ii) Ensuring that microorganisms which do get in remain as dormant as possible by providing them with poor growing conditions. This is achieved by cooling the milk quickly and to as low a temperature as possible. Cooling does not kill microorganisms but slows their growth and multiplication.

Microorganisms invariably get into milk but it is possible to keep them to a minimum. Good facilities (in the way of buildings and equipment) make the task easier but do not guarantee clean milk if the methods used are careless or slovenly. If clean, wholesome milk is to be produced, then attention must be paid to the sources of contamination and the routes by which such contamination enters the milk.

The Cow Herself

The udder contains its own natural bacterial population called udder commensals. These microorganisms are harmless in a healthy udder and contribute only a nominal 200–300 bacteria per millilitre in the bulk milk. A diseased cow or udder, in contrast, particularly if suffering from mastitis, contributes very high numbers of pathogenic microorganisms to the bulk milk. Often the addition of milk from just one mastitic cow is sufficient to raise the Total Bacterial Count of the farm bulk milk to an unsatisfactory level. Mastitic milk is also associated with higher cell counts; such milk should always be withheld from the bulk milk. Also withheld should be bloody milk, colostrum (i.e. first milk up to 72 hours after calving) and milk produced within 72 hours of a cow's last treatment with intramammary antibiotics. Foremilk—milk lying in the teat between milkings—provides ideal conditions for bacterial growth and multiplication. Prior to milking it should be drawn separately into a suitable vessel, which also affords a visual examination for mastitis, blood or other abnormalities.

Dirty udders and teats also contribute to the total contamination. Large numbers of microorganisms are always present on the skin and hair, in soil, dust, dung, bedding and fodder. For this reason satisfactory cleaning of the udder and teats prior to milking is required to clean the skin, and to remove loose hairs, bedding, soil and dung.

Dairy Workers

Natural microorganisms are always present on the skin, hair and clothes of milkers and persons handling milk. In addition hands and clothes become contaminated with soil, dung and dusty fodder or bedding. It is therefore necessary to wear clean overalls and a headcover and to wash hands frequently, particularly during milking. In addition a diseased or infected person with uncovered septic sores could transfer serious human diseases to the milk, e.g. food poisoning or gastroenteritis. Under the Milk and Dairies (General Regulations) 1959, notification of disease in respect of the farmer, his family, his workers or their families to the Medical Officer of Health is mandatory.

Equipment

Poorly cleaned and sterilised equipment is recognised as one of the major sources of contamination of milk on the farm and includes all plant surfaces in contact with milk: from teat-cup liners, to jars, pipelines and the bulk tank. Each of these presents different maintenance, cleaning and sterilising problems, and careful consideration should be given to each in terms of methods, techniques, materials used and the systems of checking efficiency of the operation.

Indicative of poor cleaning technique is the development of milk stone, a thin, yellowish scale that builds up on the surface of equipment. This visible scale consists of milk solids and deposited water salts, supporting and protecting very high numbers of microorganisms which continually contribute to the bulk milk total. Removal requires specialist treatment.

Areas of special attention for maintenance and cleaning include rubber parts, bends, agitators, valves, cocks, pipe joints and outlets. The achievement of physical cleanliness and satisfactory sterilisation of plant and equipment is so important that in certain circumstances it is worthy of calling in specialist advice and services.

Atmosphere

Microorganisms are always present in the surrounding air and capable of falling on open surfaces or being sucked into the plant under vacuum. The degree of atmospheric contamination is directly related to the conditions of the surrounding area, draughts and air flow. Milking areas are often contaminated by the movement of dusty foods or bedding, or the presence of animals which have direct and open access to polluted areas, e.g. dung heaps, slurry and silage pits, calf pens, cattle sheds or food stores.

Water Supply

Water used for drinking, cleaning and cooling must be of the highest quality both chemically and microbiologically. Hard water containing a high proportion of calcium and magnesium salts can result in the precipitation of stone scale, which interferes with the efficiency of cleaning and is responsible for the furring up of hot water pipes and heaters.

Impure water, polluted by soil, vegetation, animal dung or excrement of rodents, birds or insects contains high numbers of microorganisms and is a serious risk to milk contamination and human health. Water supplies other than mains water should be chlorinated and if stored in holding tanks should be protected by a proper covering. A sufficient and satisfactory water supply is conditional to obtaining a milk producer licence.

Cleaning Procedures

Whilst cleaning and sterilising of dairy equipment can be carried out separately or simultaneously, they are, in fact, two distinct actions. Cleaning is the complete removal from the surfaces of all milk residues, physical dirt and other deposits. The surfaces should be visibly clean under good light, have no odour or greasy feeling to the touch and should show no discolouration when the equipment is wiped with a white tissue. Very minute traces of milk residue, invisible to the eye, can support microbiological life which can multiply into large numbers on the surface. Sterilisation, on the other hand, is the destruction of micro-organisms which remain on the surface after cleaning, thereby preventing recontamination of the milk when the equipment is used again.

The efficiency of the cleaning and sterilisation process can be measured by specialist swab and/or rinse tests which count the number of microorganisms present on equipment surfaces. Calling in professional services to undertake these tests on a regular basis is a worthwhile consideration.

The basic principles and stages of cleaning and sterilising dairy equipment are as follows:

- Removal of the major milk residues from the plant and equipment by complete draining of jars, pipelines and bulk tank pumps, valves or outlets. In addition any milk filters should be removed from the system.
- Prerinsing of all surfaces with cold or tepid water as soon as possible after the milk has left the equipment. This should be continued until the rinse water is clear. Adequate prerinsing is perhaps the most important stage in cleaning. As all milk constituents, apart from fat, are soluble in cold water, they are largely removed by rinsing. Water in excess of 35°C tends to coagulate milk proteins and causes the milk sugar (lactose) to become sticky. The protein and lactose can then combine to form a deposit which when hardened becomes milk stone or scale. It is equally important for rinsing to be carried out as soon as possible to keep the milk from drying, since dried milk is much more difficult to remove.
- Washing of all surfaces with hot detergent solution. In the case of bulk tanks with ice-banks or direct expansion refrigeration coils, it is necessary to use specialist cold cleaning materials.

  The detergent is an aid to cleaning and contains a mixture of alkali salts and after chemicals with a number of useful properties.

  Some of the alkali in detergent combines with the insoluble fat, changing it to a water-soluble soap. The detergent lifts the remaining solid residues off the surface and holds them in solution or suspension allowing them to be carried away with the detergent. Added chemicals soften the salts causing hard water, improve rinsing ability and prevent the formation of scum and hard-water scale. In some Cleaning In Place (CIP) systems, heat and alkali or acid are combined in order to kill any microorganisms.

  Cleaning at this particular stage requires some physical action to assist the removal of milk and dirt deposits. Such action could be in the form of hand scrubbing, pressure sprays or, in the case of CIP, flow velocity and turbulence within a pipeline.

- Thorough rinsing of surfaces with hot or cold water in order to remove detergent residues containing salts of the detergent, milk solids and microorganisms. If detergent residues are left on equipment, they will dry, becoming hard and again forming milk stone, which requires specialist acid or proprietary treatment for removal. The presence of milk stone, therefore, indicates poor cleaning technique.
- Sterilisation of equipment surfaces to kill any remaining microorganisms. There are three main methods of sterilisation:

  Steam at 100°C, applied for a minimum of ten minutes from the time when condensate temperature reaches 80°C. This is the most effective method of killing microorganisms, but is expensive and inconvenient if the equipment has to be used while still hot. Steaming is generally used in tanks, pipelines and clusters, particularly as an occasional reinforcement of another method.

  Hot water at 80°C, applied for 10–15 minutes. This is fairly effective at killing microorganisms and is slightly less expensive than steam. In common with steam, hot water sterilisation works best with enclosed systems and CIP.

  Chemical sterilisation. This is the most commonly used method. Only chemicals approved for use on food containers are permitted. The most common sterilants are based on chlorine, iodine or quaternary ammonium compounds. Chemical sterilants are less effective than steam or hot water, and some may kill only certain microorganisms.

  If sterilisation is to be effective, equipment surfaces must be thoroughly clean and free from organic residues. Such residues protect microorganisms, waste the sterilant's oxidative properties, and generally reduce its effectiveness. In addition, chemical sterilising agents must come into contact with all surfaces for sufficient time if all the various types of microorganism present are to be killed.

  Warm sterilant solutions are generally more effective than similar cold solutions, but warm solutions containing chlorine should be avoided since they are exceptionally corrosive to plant and equipment, even if made of stainless steel.

  Detergent-sterilisers are a mixture of detergent and chemical steriliser (either chlorine or quaternary ammonium compounds) and clean and sterilise in one operation. They may be designed to operate either cold or hot. Some solutions tend to foam and are, therefore, unsuitable for circulation cleaning.

  No single detergent, chemical sterilant or detergent-steriliser is designed to cover the whole range of dairy cleaning requirements. The products used should be chosen for the particular cleaning job, e.g. cold cleaning of bulk tank; CIP cleaning of milk jars pipelines and clusters; hand washing of small utensils or cleaning of floors and drains. Care should always be taken to follow the manufacturers instructions regarding concentration, temperature and contact times.

CHAPTER 3

BUILDINGS AND EQUIPMENT

For milk production to be carried out on farms, certain items of equipment are essential. These can be classified in three categories:

- Accommodation for the cow and young stock (where rearing is undertaken) including feeding arrangement and bulk feed stores;
- Equipment for milking the cow and subsequent handling of milk;
- Fixed equipment in respect of solid and liquid manure storage, and cattle handling facilities.

TO BUILD OR NOT?

Buildings

The provision of buildings requires considerable capital but is very important for the well-being of both the dairy herd and the staff looking after the herd. Obviously economic aspects are all-important and consideration must be given to the costs of providing accommodation. For reasons of economy, modern farm buildings have moved away from traditional brick and stone structure. The modern dairy unit may well consist of pre-fabricated units often provided as part of a package deal designed to maximise the benefits accruing from economies of scale in production.

A low-cost building which is cheap initially may well require substantial maintenance work each year and have a relatively short life. A more expensive building will last longer and may require little maintenance. The prospective purchaser should consider the annual cost of a building, that is, its depreciation and annual maintenance charges, as well as insurance. However, feasibility should be considered. It is important, particularly if the investment is being financed on borrowed capital, to ensure that sufficient income will be generated to cover financial charges. Therefore there may be no option other than to consider low-cost buildings.

Adequate accommodation will be required for the milking herd including young stock. More mature stock, in-calf heifers and dry cows, can be left outside in appropriate circumstances. Accommodation will also be required for fodder, feeding arrangements, ancillary buildings and provision will be needed for waste disposal. The cost of providing access and services must not be overlooked or understated.

Quite apart from benefits in cow nutrition and improved production there are obvious benefits to management and labour utilisation where cows are housed in some kind of group.

GENERAL BUILDING LAYOUT

During the course of a winter, a cow is likely to consume something between four and ten tonnes of food and to produce four to six tonnes of dung, depending upon the intake of the animal, the productive level attained and the dry-matter content of the diet. Such large amounts of food and waste products present a considerable problem and should be considered a major factor in the planning of any layout. A high proportion of the activity on any farm takes place around the farmstead. The implications of design and layout on the cost of labour are obvious.

It is essential to have made a number of decisions before designing a layout.

1. The amount of capital available for the development.
2. The size of herd envisaged. This will depend not only on the area of land available, but also on the economic size of the unit in relation to available labour resources. This is true whether dairying is the sole enterprise or not.
3. Management policy, that is breeding policy, calving patterns, production patterns, feeding policy (hay or silage, controlled or self-feed), milking arrangements, type of housing and method of waste disposal.
4. The siting of the unit. An ideal site will have the following features:

   - Be situated in the centre of the grazing area; walking cows long distances uses up energy which would be better utilised for milk production.

(text continues on page 44)

A modern herringbone parlour with low-level jars and automatic cluster removers.

A vacuum recorder being used to test the action of the pulsators.

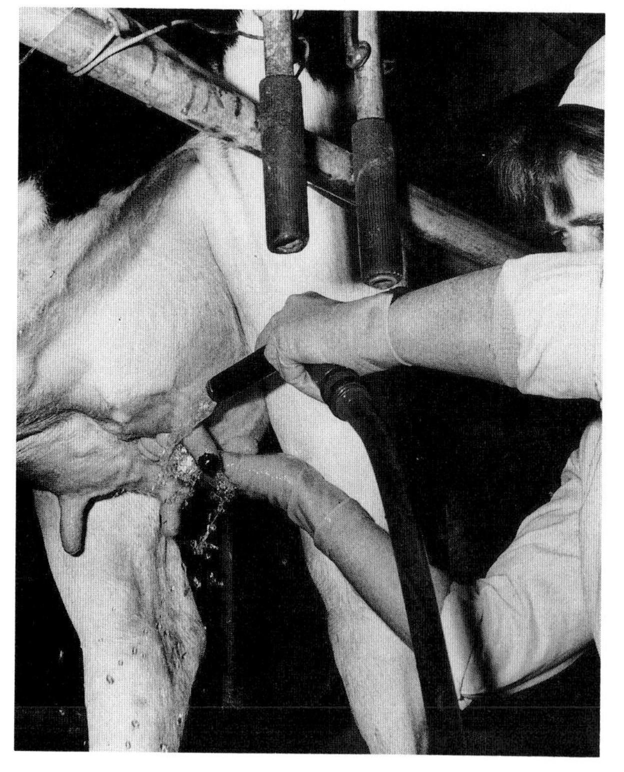

(*Above*)
These cows, on self-fed silage, are housed in winter in the cow kennel house built of wood and corrugated sheets.

Udder washing with hand spray; note rubber gloves.

This cubicle building contains 190 stalls of tubular metal construction. Partly open sides provide good ventilation and prevent roof condensation.

A simple type of cubicle installed in a shed with a wide slatted-floor passageway.

Cows in this double row of kennels have access to a 42.6 m long covered feed passage. The roof of the passage overhangs 2.4 m on either side, enabling the cows to eat in comfort whatever the weather.

Forage feeder in action which is capable of discharging either side.
Kidd Farm Machinery Ltd.

A block cutter helps in an easy-feed situation.
Parmiter and Son Ltd.

Strawed yard in a Crendon wide-span concrete building.

Cows at feed face in typical package-deal building at NAC, Stoneleigh.
Farm Plan Construction Ltd.

Automatic slurry scraping system.
Alfa-Laval.

(Above)
In this NIRD-designed rotary herringbone layout the partly-opened rump gate guides the entering cow. A sheeted gate covers the manger of the stall through which the cow walks to enter the second stall.

(Right)
Slatted floors can increase stock density and reduce capital investment but demand good stockmanship.

- Have good access. More and more materials are handled in bulk requiring the use of large vehicles to make deliveries. The size and turning circles of these vehicles must be allowed for.
- Be well serviced. If not already available, water and electricity should be obtainable at reasonable cost.
- Be well drained, preferably elevated.
- Be sheltered; the site should not be too exposed and the direction of prevailing winds should be borne in mind.
- Have site conditions which are suitable for the structural requirements of the buildings. Made-up ground, peat, etc., must be avoided.

The main features of any unit are as follows:

- The cows must be housed in comfort with adequate light and ventilation. Watering facilities are vital. A high-yielding cow requires 40–50 litres per day, the quantity being affected by the dry-matter content of the ration. Restriction of water intake will depress yields.
- The accommodation should be planned to bring together housing, feeding and milking operations, so that staff time is used to its best advantage. Modern framed buildings can incorporate large spans which bring together these functions under one roof. A stockman must have adequate time to deal with his stock, and he will not do this if he spends his time rushing around a badly laid out set of buildings.
- It should be well sited as described.
- It must comply with the necessary statutory requirements; certain parts of the premises must conform, for example, with the standards of the Milk and Dairies Regulations, 1959.
- Attention must be given to the siting of bulk food stores, silage clamps, etc. Ideally all should be under one roof; however, cost may preclude this. Modern silage-making practice involves covering the clamp with a sheet; thus a roof is by no means essential—its main benefit is likely to be to the stock. The capital cost of a roof may well be better invested in the provision of a feed area.
- The buildings must be designed and laid out to avoid pollution from slurry, silage effluent, dirty yard water run-off and waste milk or dairy washings. Cow movements across open yard areas need to be kept to a minimum to prevent fouling and increasing the volume of dirty water.

- Young stock accommodation, calf pens, AI stalls, bull pens, isolation boxes, cattle handling facilities, office and staff accommodation should be provided on the same site. Often older traditional buildings can be adapted to these uses.
- The dairy must be adjacent to the milking parlour and have good access for the road tanker.
- The design should allow for future expansion, however unlikely, and suitability of the buildings for other uses should be considered.

COW HOUSING

Traditionally, cows were housed in cowsheds and tied by the neck, using chains, leather straps or metal yokes. After the Second World War the increase in labour costs provoked a move from cowsheds and the introduction of the loose housing system on straw. During the 1960s the yard and parlour system was widely installed. Bucket milking and latterly pipeline milking were increasingly replaced by the modern milking parlour.

Loose Housing

This system gives a greater flexibility of cow numbers linked to the increase in herd size and the need for speeding up the milking process.

One disadvantage of loose housing associated with increasing cow numbers was the loss of the opportunity for individual attention to cows. This feature was to some extent minimised by the feeding systems which operated through the sixties. Large cow numbers were associated at this time with favoured low input/low output systems, concerned with the balance that could be achieved between silage and concentrate.

Advantages of the yard-and-parlour system over the traditional cowshed housing and milking are:

1. The daily requirements of cleaning out the cowshed to meet Milk and Dairies Regulations are reduced to littering the sleeping area of the yard and scraping the loafing and feeding area only. The strawed area builds up to a deep bed of farmyard manure by the end of the winter period and this can be easily handled mechanically in a very short space of time.
2. During the sixties, there was a move towards more intensive

Example of a fixed herring-bone parlour. In this installation on a Somerset farm there are 10 units and 20 stalls (10/20).

A new prefabricated herringbone parlour with its circular collection/dispersal yards. The squeeze gate is weight-operated.

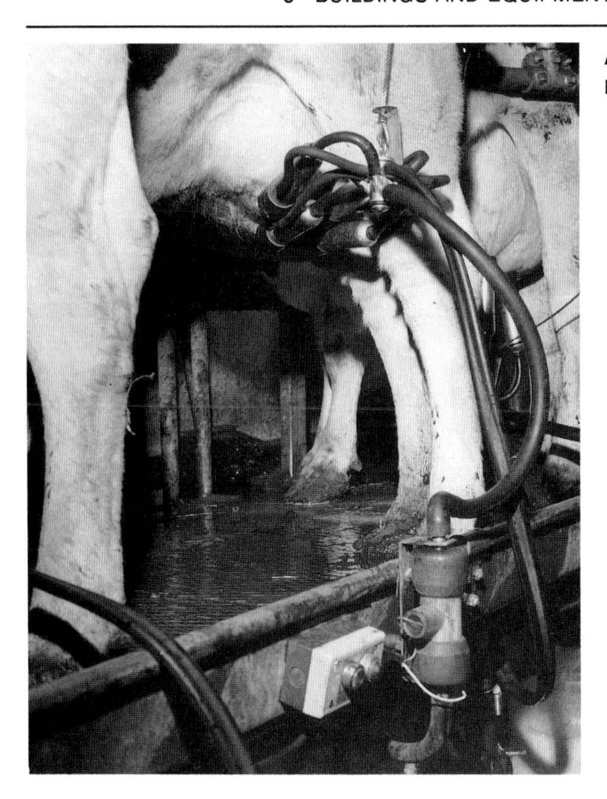

Automatic cluster removal in action.

grassland management, leading to higher stocking rates and greater cow numbers, which could be more easily accommodated by the yard system.

3. Self-feed silage was another product of the improving grass technology and this also fitted in with loose housing. Individual silage feeding is too heavy a task to be contemplated by farmers using a cowshed system.
4. The parlour milking system has considerable advantages over the cowshed bucket milking system; the introduction of circulation cleaning and bulk tanks has brought about greatly improved labour use.

Yard Layout

A yard area should consist of a lying or bedded area and a loafing or feeding area.

The space required for each animal depends upon the age and size of stock, the bedded area and feeding system. For adult cows an allowance of 4.2 m^2 for lying area and 2.0 m^2 for loafing and feeding is typical. The feeding area can give access to a feed face or self-feed silage area.

The sleeping area must be covered. A feed fence can be situated along the outside of the yard so that feeding can be done by mechanical means such as forage boxes. All feeds and manure should be handled as little as possible over the shortest possible distance.

Slatted floors may be used in the loafing area; normally 2.8–3.4 m^2 per head should be allowed, depending upon the breed. Slats are usually of concrete up to 3.5 m in length; depth, depending upon span, 100–225 mm; top width 100–225 mm tapered, so that the base is narrower than the top. The recommended width between slats is 25–40 mm. Slatted floor systems reduce bedding and labour cost but are more expensive initially. Welfare codes preclude the use of slats over the entire area.

Cubicle Housing

The most widespread development in loose housing was the introduction of cow cubicles in a covered yard, and cow kennels where only the lying area is covered. Kennels form a very low-cost housing system. Cubicles give substantial savings in bedding costs and the cows are cleaner, consequently saving washing time and speeding throughput in the milking parlour.

The cubicle should be comfortable and afford the cow protection against disturbance and injury from its neighbours. Various designs of cubicle are now in widespread use. If sufficient regard is not given to the needs of the animal, however, problems will arise, not only with disease and injury but also with cows refusing to use the cubicles.

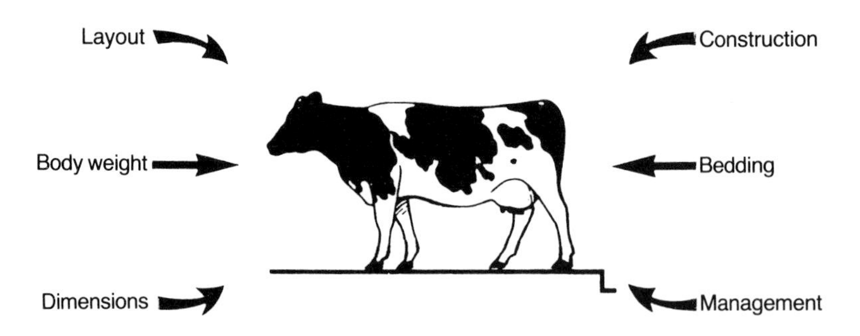

Figure 3.1 Design factors for cubicles
MAFF Booklet 2432

Hosier-Guery's 270 litre glass reinforced plastic mini-vat has a stainless milk vessel and water jacket.

Bulk milk production. As the tank is not kept under pressure, a weight-operated valve is necessary. When the containers hold a certain quantity of milk, it switches on the motor and cuts out the vacuum so that the milk can be piped into tanks, each of which holds 2455 litres.

Slurry separation.

This scraper has a 1.84 m wide chain blade fitted with a reversible rubber squeegee on its base. Side plates are toed outwards to give 2.13 m sweep and their 0.61 m reach can be set from the tractor seat in a forward position (for pulling) or a backward position (for pushing).

Length—the normal length of a cubicle is 2.13 m but this should be related to size of animal in the herd. A head rail or brisket board may be necessary to prevent fouling of beds by those cows lying too far forward in the cubicle.

Width—this is often varied to fit the appropriate bay sizes. The width should be 1.2 m but this can be reduced for smaller cows.

Heelstone—this consists of a step 200–250 mm above the dung passage. The cow standing on the heelstone should dung directly into the passage, thus keeping the bed clean.

Cubicle Divisions—various types have been developed. Most designs have two rails along the length. The height of the lower rail is critical; it should be 400–450 mm above the base. If this measurement is not correct, the animal may be injured or easily jam itself underneath the rail. In some designs a nylon rope is used in lieu of a bottom rail.

The 'Dutch Comfort' cubicle division, which has no horizontal supports, is designed to allow the animal to share its neighbours' space which assists rising and increases comfort.

Beds—a slope of 75–100 mm is recommended as this helps the drainage. Bases have always posed a problem, because cows prefer a soft comfortable lie.

Box muck, soil and rammed chalk are all suitable materials for cubicle beds, but require maintenance and can also provide an ideal environment for microorganisms. Concrete and bitumen macadam beds are more durable. Concrete should ideally be damp-proofed and can also be insulated. If bitumen macadam is used it must be of the 'dense' type. Bedding is still necessary with both these alternatives.

Cow mats made from a number of materials are becoming more popular. Some bedding is still required, though the quantity is reduced.

Passages—the dung channel between heelstones should be not less than 2.4 m wide to allow for cow movement. It may have to be varied, particularly if the cubicles are being fitted into existing buildings. The channel also needs to be wide enough to allow access for the slurry scraper and should be carefully considered, as narrow passages lead to more slurry. The surface should be non-slip. Slats can be used which reduce the labour costs, as do automatic scrapers. However, provision should be made for access by tanker or irrigation equipment to remove the slurry from tanks beneath the slots.

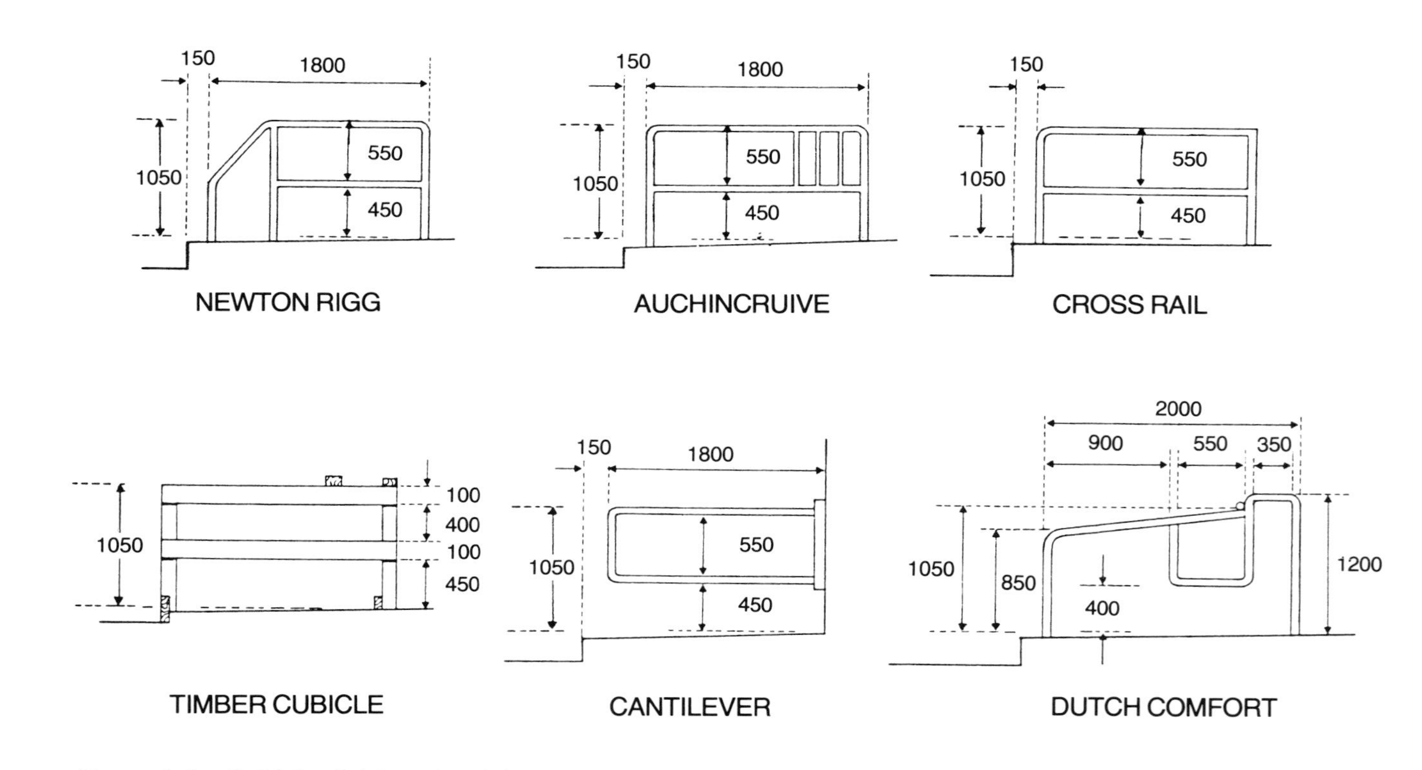

Figure 3.2 Cubicle divisions for dairy cows
MAFF Booklet 2432

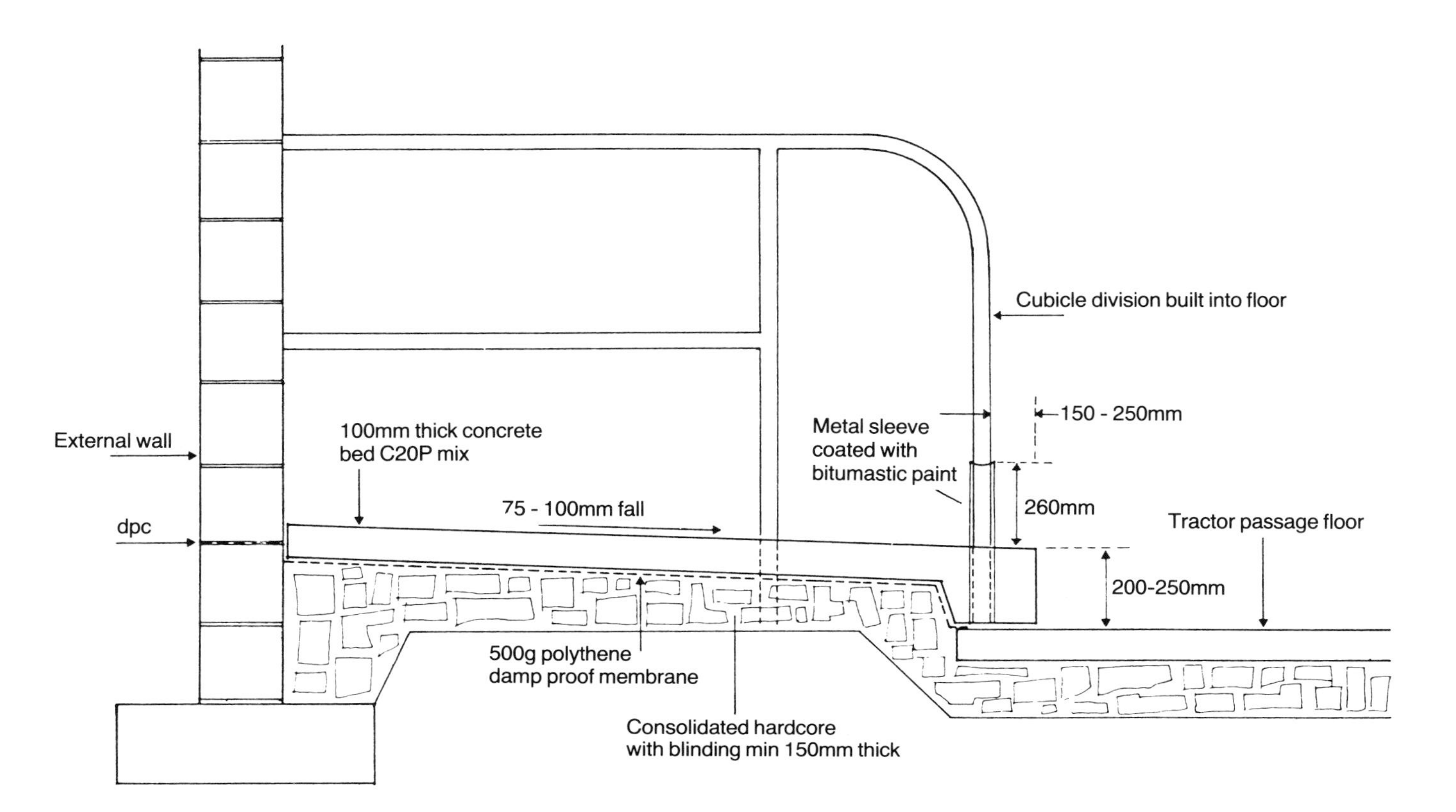

Figure 3.3 Cubicle with concrete base and in situ kerb
MAFF Booklet 2432

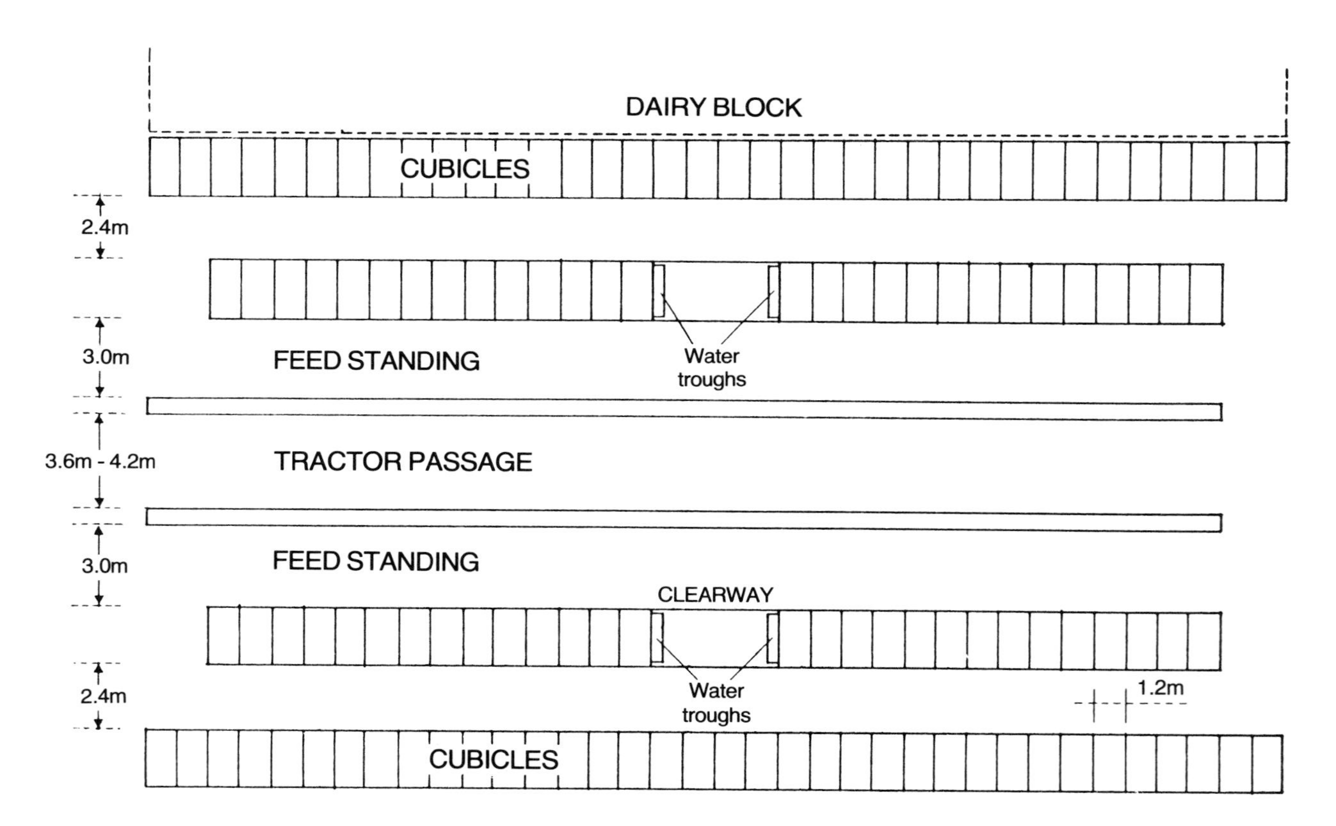

Figure 3.4 Cubicles with controlled feeding for 128 cows
MAFF Booklet 2432

Feeding Arrangements

The feed area should be adjacent to the housing to reduce difficulties of cow movements and fouled areas. The feed area should be roofed, although this is a costly item.

Self-feed silage is of course the simplest system. The depth of the silage face should not exceed 1.83 m and if there is 24-hour access, 18–23 cm of the face should be allowed per cow, depending on the breed. If the silo is not roofed, however, then rainwater will add to the volume of slurry causing increased storage and disposal problems along with increased risk of pollution.

Feed passages should normally be double sided to utilise space to the full, with 4.57 m between feed fences to allow access for forage wagons or the use of a block cutter. If foods are ad lib, 30 cm per cow should be allowed, but if access is restricted, allow 60 cm. There are various designs of feed fence, from single horizontal rails to diagonal barriers. Diagonal barriers help reduce bullying and waste. The provision of a feed fence allows for the use of other bulk foods and supplementary feeding of concentrate.

Milking Arrangements

In machine milking equipment, there are four distinct types of plant:

1. The bucket type with a tapped vacuum line fixed overhead, installed in a cowshed.
2. Pipeline milking, involving a milk pipeline into which the milk is delivered direct from the teat cups, without intervening buckets. Recording is done by the use of a milk-flow meter of which there are several approved models available. A long pipeline should have as few vertical lifts and right-angle bends as possible since these harbour milk residues which may develop taints as a result of breakdown of milk fat.
3. The fixed milking parlour or movable bail for use with yarded cows or cows at pasture.
4. The rotary system in which cows circulate on a revolving platform past the operator.

The principles involved in the parlour and bail systems are similar: by centralising milking at one point, the operator's time is saved, as the cow does the walking. The operator concentrates on washing and milking the cows in relay and probably feeds the concentrate ration to each cow—all with a minimum amount of movement on his part.

Conveyance of the milk from the milking point is done by vacuum.

MILKING PARLOURS

The design of milking parlours and equipment has made rapid strides over the years. The main types of milking parlour are as follows:

1. Abreast

The cows enter through the operator area and leave through the front of the stall. The standings are raised 350–400 mm above the floor level of the operator. More attention can be paid to individual cows as they enter and leave singly. The amount of walking involved for the operator is the limiting factor on the size of this type, thus making it more suitable for the smaller herd.

- *Simple Abreast*—Cows stand in line abreast. Parlours larger than six stalls involve too much walking for the herdsman.
- *Circular Abreast*—The cows stand in single stalls forming a circle round the operator pit in the centre. Walking is reduced, but the presence of cows in the working area poses a disadvantage.
- *Back-to-Back Abreast*—As its name suggests, the cows stand at either side of the operator pit in single stalls.

Both the circular and back-to-back allow the use of more stalls and allow the parlour to cope with larger herds.

2. Tandem

The cows stand-on in line to the operator's pit between two rows of stalls. Each has separate side entry and exit gates. Again the amount of walking involved is a problem and a relatively large building is required to accommodate a small parlour. This system is no longer widely used.

3. The Chute

This is a modification of the tandem particularly suitable for narrow buildings in which cows enter and leave in batches.

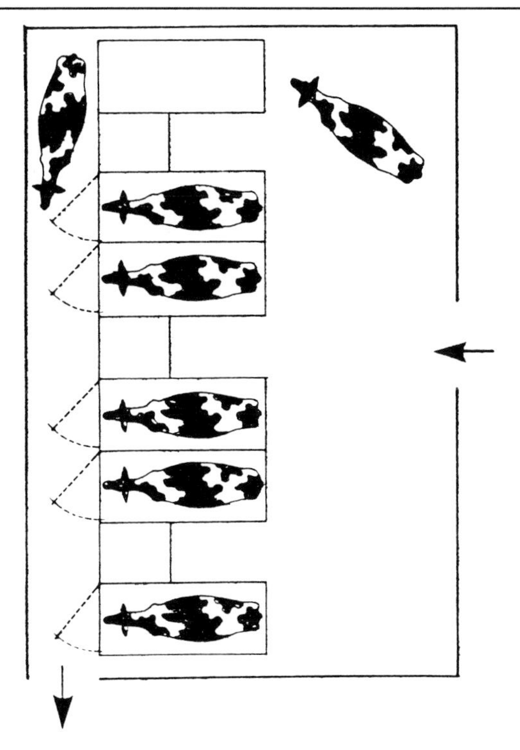

Figure 3.5 The six-stall, three-unit abreast parlour
MAFF Booklet 2426

For large herds, though, the length of the parlour becomes unmanageable.

4. Herringbone

The cows stand in groups at an angle of 30–35 degrees on either side of the central operator pit. The average pit depth is about 0.9 m, the width will depend upon design, for example, a distance of 1.8 m is allowed for a parlour with single row, eye level jars. The herringbone substantially reduces the amount of walking, and relatively large parlours can be installed in short buildings. This type of parlour can be easily automated—by the installation of mechanical gates, automatic feeding and cluster removal—to make it more suitable for the larger herd.

5. Polygons and Trigons

These are really in essence multi-sided herringbones.

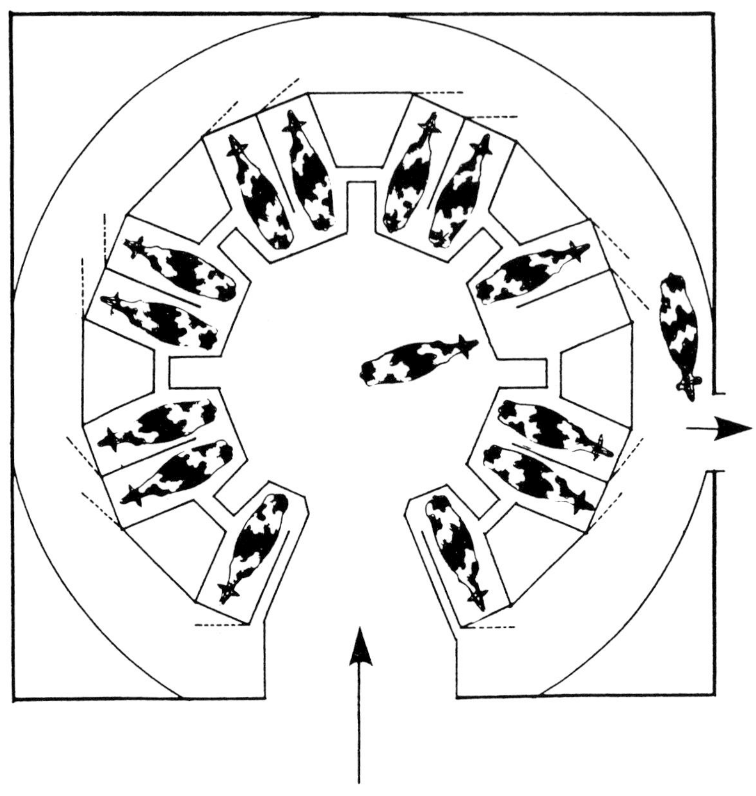

Figure 3.6 The circular abreast parlour
MAFF Booklet 2426

6. Side-by-Side

Similar to the style of the herringbone but the cows stand at an angle of 90 degrees to the pit. This reduces the length of the building required and allows more efficient automatic cluster removal.

7. Rotary

The parlours rotate round or past the operator and incorporate either the tandem, herringbone or abreast principle. The complete rotation may take six to seven minutes, however, movement can be halted while cows move into and out of their stalls. Greater throughput is claimed but such parlours are more expensive, can take up more building space and are obviously more prone to mechanical breakdown.

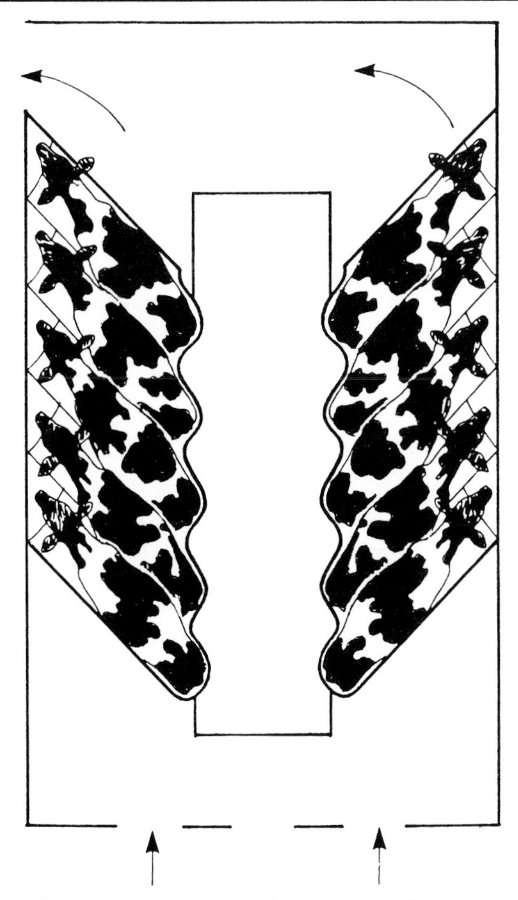

Figure 3.7 The herringbone parlour
MAFF Booklet 2426

The interest in rotaries has waned with an increase in the larger herringbones.

Two Stalls or One per Unit?

The main advantage of two stalls per unit is that the operator can apply the units in the order he wishes, thus allowing for the cows taking longer to milk. Throughput is increased and the centre work area is freer. Although throughput is not doubled, it can be speeded up more effectively by increasing the length of the parlour rather than doubling up where this is possible. Doubling up substantially increases running costs. Certain parlours, that

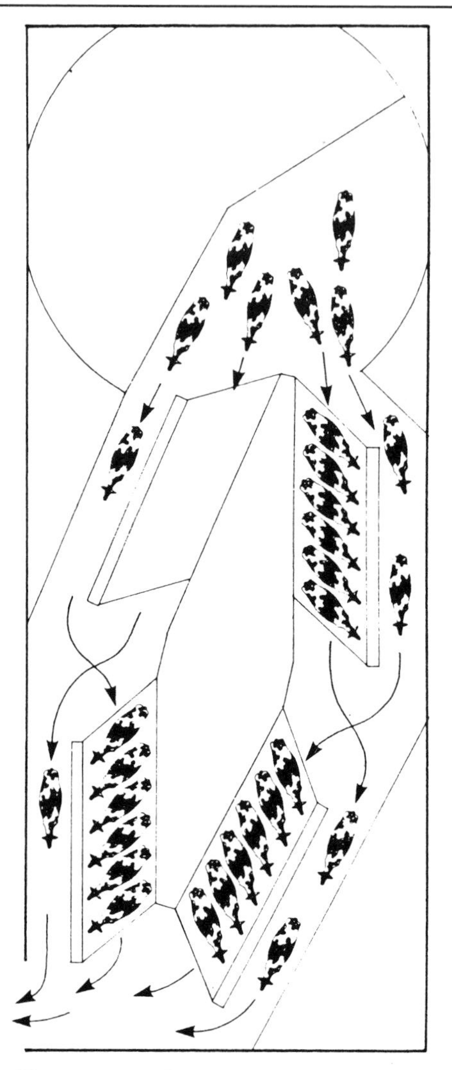

Figure 3.8 The polygon
MAFF Booklet 2426

is, rotaries, polygons and trigons, are 1/1 of necessity (that is one per unit).

Interest has been focused on the idea of dispensing with jars and milking direct to the pipeline. This would improve the working area, but would create less vacuum reserve and increase the risk of milk contamination.

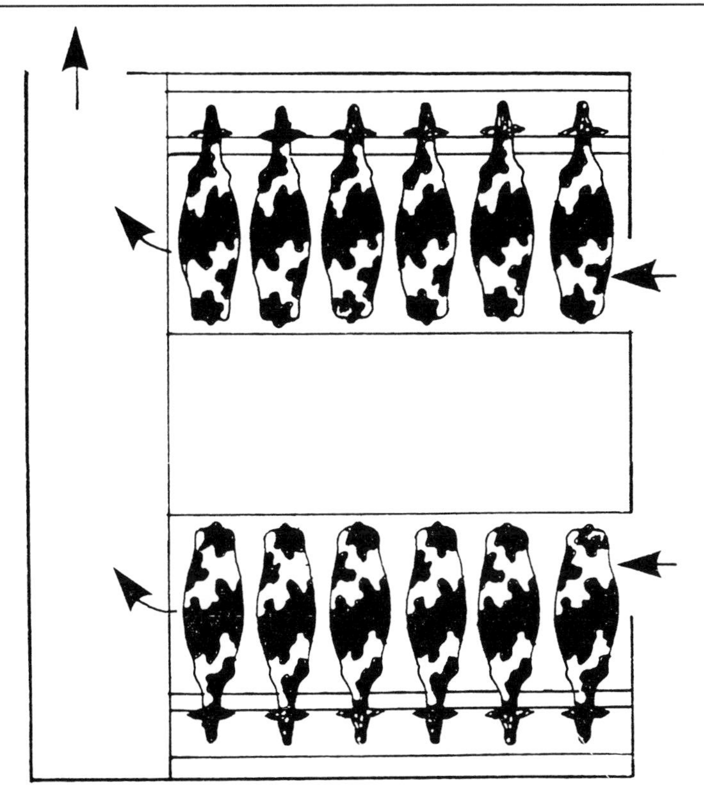

Figure 3.9 The side-by-side design
MAFF Booklet 2426

BS 5545 (1980). Milking Machine Installations

All new installations or alterations to existing parlours should comply with the specifications of the BS 5545 (1980) for the design, installation and mechanical performance.

Before installing a parlour, the following factors should be considered:

1. Feeding of Concentrate

Two stalls per milking unit will usually give sufficient time for the cows to eat concentrates if they are fed wholly in the parlour. Cows vary in the rate at which they eat; this is usually 0.5 kg per minute for cubes, but slower for meal. Slow eaters or feeding of

Figure 3.10 The rotary abreast parlour
MAFF Booklet 2426

excessive concentrates will slow down the milking process. The use of a feed fence outside the parlour alleviates this problem.

2. Size of Herd

If the total time spent on milking is not to exceed one and a half hours in the evening and two hours in the morning, then there should be between ten and fifteen cows per unit, depending upon the layout, yield per cow and the spread of calving pattern.

The number of units a man can handle without being overtaxed or underemployed is largely governed by what he has to do in

Table 3.1 Work Routine Time

Operation	*min/cow*
Let in and feed cow	0.15
Foremilk	0.10
Wash and dry udder	0.20
Attach cluster	0.20
Remove cluster	0.10
Disinfect teats	0.10
Let out cow	0.10
Miscellaneous	0.05
Total	1.00

Source: MAFF Booklet 2426.

milking each cow—the so-called work routine. A typical work routine is shown in Table 3.1.

The potential throughput of a parlour depends on the milking-out time, the number of units and the length of the work routine. Therefore, throughput can be improved by mechanisation and automation of certain operations, for example, mechanical entry and exit gates, concentrate feeding with the aid of electronics, teat disinfection and washing sprays, milk recording and automatic cluster removal (ACR). The main benefit of ACR may however be to prevent overmilking rather than to save time. Table 3.2 shows the effect of automation.

In the case of low yielding cows the limiting factor will be work routine; for high yielding cows it will be milking-out time which allows ample time for full work routine.

3. Available Buildings

The third consideration is whether an existing building can be adapted to house the chosen parlour or whether to build afresh. Whatever the decision it is important to take account of the need to site the parlour properly so as to allow easy passage of cows to and from it to the cow housing.

Dairies

All milk is collected in bulk from farms in England and Wales.

A refrigerated milk vat (tank) must be installed and therefore the dairy must be large enough to accommodate this while leaving sufficient headroom to allow calibration. In determining the size of the tank, allowance should be made for possible herd expansion or increased yields per cow.

At the time of collection the tanker driver inspects the milk in the tank, checks the temperature, agitates the milk and takes a sample for testing under the various schemes controlling milk quality. The quantity is measured by either a dipstick and calibration chart or a flowmeter.

Table 3.2 Effect of Automation

Operation				*Man minutes/cow*			
Let out cow	0.20	0.10	0.10	0.10	0.10	AUTO	AUTO
Let in and feed	0.25	0.15	0.05	0.05	0.05	0.05	AUTO
Foremilk	0.10	0.10	0.10	0.10	0.10	0.10	0.10
Wash and dry teats	0.20	0.20	0.20	0.20	0.20	0.20	0.20
Attach cluster	0.20	0.20	0.20	0.20	0.20	0.20	0.20
Remove cluster	0.10	0.10	0.10	AUTO	AUTO	AUTO	AUTO
Disinfect teats	0.10	0.10	0.10	0.10	AUTO	AUTO	AUTO
Miscellaneous	0.05	0.05	0.05	0.05	0.05	0.05	0.05
TOTAL	1.20	1.00	0.90	0.80	0.70	0.60	0.55
Cows/man hour	50	60	65	75	85	100	110

1. Manual
2. Mechanised barriers/gates
3. Semi-automatic feed dispensers
4. Automatic cluster removers
5. Automatic teat disinfection
6. Automatic cow exit
7. Automatic cow entry

Source: MAFF Booklet 2426.

Table 3.3 Types and Sizes of Parlour for Varying Herd Sizes

Size of herd (cows)	*Maximum milking time (hours)*	*Required throughput (cows/hr)*	*Number of operators*	*Size and type of parlour*
50	1.5	34	1	6/3 abreast
80	2	40	1	10/5 herringbone
120	2	60	1	12/12 herringbone
150	2	75	1	16/16 automated herringbone
250	2.5	100	1	16/16 trigon
350	3	117	1	20/20 rotary

Source: MAFF Booklet 2426.

Collecting Yard

The collecting yard holds the cows prior to entry to the adjoining milking parlour. It is preferable that the cows go straight into the parlour without turning corners. If the parlour and collecting yard are continuous without a dividing wall, this will encourage cows to enter. This does mean, however, that the collecting yard must be cleaned to the same standard as the Milk and Dairies Regulations require in the parlour. Allow 1.2–1.5 m^2 per cow depending upon the breed, the yard being preferably long and narrow.

A diversionary pen at the exit from the parlour allows cows to be held for veterinary treatment or AI on leaving the parlour. Ideally it should be possible to operate the gate from the parlour pit.

A foot bath with a minimum depth 150–200 mm of liquid in two sections, each 2.13 m × 1.22 m, one for washing and one for disinfecting, should be incorporated in the exit races.

Waste Disposal

In the earlier housing methods it was usual for the dung to be removed from cow houses on a regular daily basis. With yarded systems and mechanisation it is more likely that the bedded area will be cleared out once a year, usually by contract. On cubicle systems, particularly where cows are silage-fed, slurry is the inevitable result. This is usually scraped out daily by either tractor-mounted or automatic scrapers. The automatic systems now available are usually very reliable, and are normally operated by a chain or wire rope which can substantially reduce the very high labour requirement of this task. When designing layouts the scraping runs must be carefully considered, the fewer corners the better.

Slurry is normally pushed into a store which can range from an earth lagoon to a slurry tower; only on farms with free-draining soils can slurry be disposed of daily. It is more usual for the store to be emptied at leisure when other farm jobs permit, and is often carried out on a contract basis.

With the energy crisis and the increase in fertiliser prices, it is increasingly important to realise the value of the plant nutrients which slurry contains. However, the value is not sufficient to justify any substantial capital investment in storage facilities: 10 m^3 of slurry would yield 35 kg nitrogen., 11 kg phosphate and 58 kg potash. The average dairy cow produces approximately 42 litres (0.042 m^3) of slurry per day.

Environmental concerns are an increasing feature of daily life. For livestock farmers the need to handle and dispose of wastes is an ever-present problem. Close monitoring of water quality standards in rivers, water bodies, watercourses, etc. has revealed a deteriorating situation with an increasing number of incidents caused by farm pollution. The most frequent causes of such incidents are silage effluent and slurry, whilst dirty water run off from yards and concrete aprons is also a major problem. Careless disposal of waste milk or dairy chemicals is less frequent but is no less serious when it causes contamination of water supplies.

Good design and practice will reduce pollution risks but the National Rivers Authority, which is responsible for maintaining water quality, takes the view that all agricultural pollution is avoidable. Increasingly farmers are likely to run the risk of prosecution for causing pollution, with the attendant prospect of heavy fines.

Unacceptable levels of nitrates in drinking water supplies are a growing problem; applications of organic manures such as cow slurry are a contributory factor. It may be that in the future, slurry or manure applications in autumn or winter will not be acceptable since the nutrients applied cannot be taken up by arable or grassland crops. Increased storage facilities to permit lengthy storage periods may be needed, as is common in some Continental countries.

Slurry separation is an option which is being considered by more and more farmers, either mechanically or by strainer barriers or weeping walls. Liquid effluent can be pumped to a separate store or directly onto land through an organic irrigation system. The latter is becoming more common but is really an option only on suitable soil types.

Slurry cannot be applied to the land indiscriminately because of the risk of run off and subsequent pollution. The Code of Good Agricultural practice should, therefore, be followed at all times. There is also a health risk involved and grassland should not be utilised for at least a month after slurry application, to allow disease organisms to die off, in particular, salmonella.

Calf and Young Stock Housing

Calf housing should provide warm and draught-free conditions. As hygiene is essential, surfaces and equipment should be easy to clean and disinfect. The building must be properly ventilated.

Calves can be penned individually or in groups. Individual pens should measure 1.2 m × 1.8 m incorporating solid pen divisions. The calves are fed from buckets attached to the pen fronts which may include yokes. These buckets must be cleaned regularly. Hay is fed in racks or nets which should be moveable. Alternatively, calves may be penned in groups and fed from an automatic machine. The size of the group depends upon the capacity of the machine used, but can be as many as 20. Individual pens are the more popular system.

Young cattle below twelve months of age require winter accommodation in the form of a yard either semi or fully covered, with adequate trough and rack room, that is, approximately, 45–55 cm per animal.

It is advisable to erect temporary divisions in such yards to allow calves to be grouped according to size, thus reducing bullying. Such yards need not be elaborate in construction so long as they provide a dry bed, and there is a minimum allowance of 4 m^2 per animal.

Bulling heifers and in-calf heifers may be out-wintered on many farms; if they are in-wintered, yarding accommodation can be as described above but allowing 5.6 m^2 per beast.

Cattle Handling Facilities

Many modern dairy units lack adequate handling facilities necessary for inspection and veterinary purposes. The minimum requirements are a crush and a race.

Crush—this is a pen which accommodates a single beast with a yoke to secure the head and a rump bar to secure the rear. Sideways movement should be prevented. The crush should be the correct dimensions to suit the size of stock using it; it therefore needs to be adjustable.

Race—this is designed to get cattle into single file before entering the crush. It should ideally be 75 cm wide, consisting of stout parallel fences with rails on the inside. it can be used for simple treatments or inspections. A foot bath can be provided in the floor of the race.

The race and crush can be planned for use in conjunction with the collecting yard by-passing the parlour. There should, however, be access to both ends of the crush.

Calves and Isolation Boxes

These are required for isolation of cattle for disease control, treatment or calving. They need to be sited away from the main block of buildings, and each box should be 4.2 m × 3.6 m × 2.6 m high. Ideally one box should be provided for every 20 cows, but in larger herds this can be extended to 50.

STATUTORY REQUIREMENTS

It should be noted that before work commences on the construction of new buildings, or alterations to old, certain consents may be necessary. Although development for agricultural purposes is generally considered 'permitted development', planning permission is required for certain work.

New buildings for livestock housing with the provision of slurry storage facilities will definitely need planning consent if they are within 400 m of a *protected* building. A protected building is a dwelling (not a farmhouse), school, etc. The 400 m is measured from the curtilage, i.e. the garden wall, fence etc. A change of use to livestock accommodation or slurry storage will also require consent if the building is less than five years old. All buildings must comply with National Building Regulations, but for agriculture there is some exemption.

If disposal of foul drainage is involved then the consent of the National Rivers Authority and/or the local authority may be required. Work may also need to satisfy the requirements of the Milk and Dairies Regulations where appropriate and also the Health and Safety at Work Act 1974.

BUILDING STANDARDS

British Standard 5502

This is the code of practice for the Design of Buildings and Structure for Agriculture. It:

- brings together in a single standard a large amount of readily available information;
- gives recommendations for the design, construction and provision of services to agricultural buildings;
- is performance based in designing to a particular site;
- recognises four classes of buildings in terms of design life, period and density of human occupancy and proximity to highways and dwellings;
- is available in three parts from the British Standards Institution.

In cases of doubt, professional advice may be required, but if a contractor is used he may well be familiar with the requirements of BS 5502.

Grant Aid

Grant aid is no longer available for water at facilities to be used for production purposes, i.e. cattle housing, etc. The emphasis has changed to encouraging environmental improvements by combating pollution and conserving the countryside. Grant may be available for the provision of disposal and storage facilities for dairy water, repairs to traditional buildings, environmental projects etc., under the provisions of the Farm and Conservation Grant Scheme. The nearest Divisonal Office of the relevant Agriculture Department will supply details.

Costs

A typical cubicle house with a central feeding passage wide enough to take a tractor, including internal fittings and services, would cost £600–£700 per cow place. A building to house a milking parlour, herringbone or rotary, would cost £7,000–£11,500 excluding equipment.

The parlour equipment to go into the building would range from £1,200–£2,000 per point, according to size and degree of sophistication. Automatic cluster removal would add approximately a further £1,000 per point. The dairy could be expected to cost £150 per m^2 with the bulk tank costing approximately £4–£4.50 per litre.

An above ground steel slurry store would cost approximately £20 per m^3 and the reception pit to this about £3,000.

For a silage clamp with concrete base and retaining walls, allow £30–£40 per tonne of silage; if a roof is provided, then a simple framed building will cost approximately £55–£75 per m^2 depending on size. An allowance should also be made for the provision of an effluent tank.

The Importance of Getting Advice

Erection of buildings is expensive and mistakes can be costly; lines on plans can easily be rubbed out, walls not quite so easily. The question of whether to seek advice should be considered. Professional advice can be extremely cost effective in dealing with all aspects of design, layout and construction and also for inspecting work in progress. It is important to ensure that the work is done to the required standard, that statutory requirements have been met and that it is done in the most cost effective manner.

CHAPTER 4

BASIC ORGANISATION OF A DAIRY FARM

DEVELOPMENT OF DAIRYING

In the British Isles, dairy farming first evolved as a result of the demand for butter and cheese; items which made up a large proportion of the population's diet. Cows calved naturally in the Spring and the milk subsequently produced in the summer was processed into cheese which was allowed to mature before being sold. This type of milk production was particularly common in the grassland areas of the west of Britain.

As industrial areas developed, the demand for fresh milk additional to the trade for butter and cheese also increased. Town dairies were established, stocked with black and white cattle from Holland; the local shorthorn cattle not being suitable for the higher levels of production needed by the market.

As road and rail transport developed, the town dairies were phased out and replaced by specialised dairy farms which were set up on the periphery of the areas of high population.

The introduction of the Milk Marketing Boards led to a greater use of transport facilities, allowing milk to be taken to any appropriate market, be it for liquid use or manufacturing, irrespective of where in the country it was produced.

The growth of dairy farming was also aided by the availability of imported feeding stuffs for at least fifty years prior to the Second World War. Farms were by no means self-sufficient; the home-produced maintenance foods, e.g. hay, roots and straw, were easily and cheaply supplemented by imported cereals and oil cakes from overseas.

Thus, dairy farming in this country developed largely as a processing industry. The dairy farmer was concerned not so much with growing and feeding his own crops as with the conversion of purchased feeding stuffs into milk. Only by such a system could dairy farming have survived the bleak conditions of East Lancashire, the West Riding or Lanarkshire.

Small farms in these areas produced large quantities of milk from the use of cheap imported feeding stuffs. The system did provide a means of building up fertility on these farms which would have been impossible if they had to be dependent solely on their own resources. The system which was evolved concentrated very much on wholesale milk production; few herd replacements were reared as there was a plentiful supply of down-calving heifers from more remote livestock rearing areas. This availability of replacements also encouraged farmers to sell off cows at the end of a lactation for beef, thus providing another cheap source of food for the industrial areas, particularly in the North of England.

Descriptive phases, such as 'Flying Herd' and 'Milk and Feed' are still used today in these traditional dairying areas. During and since the Second World War the pendulum has swung in the opposite direction. During the war, availability of imported feeds was greatly restricted, giving great incentive to self-sufficiency on the dairy farm. In more recent years prices of imported feeds have risen steeply and this further emphasises the continuing need for increasing self-sufficiency on the dairy farm. In practical terms this necessitates a much greater exploitation of the farm resources available. All dairy farms have a similar range of basic resources—cows, grassland, buildings and labour.

Successes in dairying will only be achieved by blending these resources to achieve the maximum profit available. There is a different recipe for success for every farm. On any given farm, the climate, the area of land available and its inherent fertility play a considerable part in deciding which system of dairy farming is to be followed.

A factor which contributed to the spread of dairying for liquid milk production was its relatively greater profitability than butter or cheese production, although these had, as ancillary enterprises, the rearing of stock and pig fattening.

Butter and cheese prices are subject to world competition. The relative freedom of fresh milk from this competition together with the work of the marketing boards encouraged milk consumption. Milk sold on the liquid market in the United Kingdom always commanded a much better price than that available from milk used for manufacturing purposes.

Many marginal farms, which had previously been concerned with sheep and beef production, changed over to milk production because of its greater profit potential and the increased stability in the form of 'the monthly milk cheque'.

SIZE OF DAIRYING ENTERPRISE

To illustrate the general organisation of dairy farming, examples will be shown of small, medium and large farms to show how land is the dominant factor in deciding the size of the dairying enterprise and the part it plays in the economy of the farm.

1. Small Dairy Farms: 10–40 Hectares

Farms of this type are generally found in lowland areas of the western part of Britain. The good fertility and fairly high levels of rainfall, combine to offer a high potential for growing grass. There is much competition for land, leading to high prices. This inevitably results in the need for high stocking rates, i.e. 2–2.5 cows per hectare.

With the increasing emphasis on grass and grass products these levels of stocking must be related to a fairly high level of nitrogen application. The higher stocking rates, coupled with the necessary higher levels of fertiliser, particularly nitrogen, inevitably lead to a need for the reorganisation of the type of winter forage used. On most farms this has resulted in a changeover from hay to silage. Bearing in mind that the basic factors of production on any farm are land, labour and capital, the shortage of land can only result in the farm necessarily becoming much more labour- and capital-intensive. Labour is often provided on the small farm by the family unit so, if production is to be raised then capital must be invested in silage making equipment. Numerous schemes have been available for grants towards the cost of providing both machinery and suitable buildings to enable small farmers to move into this area of intensive grass technology.

A major reason which persuaded many farmers into this policy was of course the vulnerability of small farms to the vagaries of the climate when they were dependent upon hay as the main fodder crop for winter feeding of cows. Barn-dried hay is made on some farms but still does not provide a safeguard against unpredictable weather.

The emphasis in small-scale farming must be on achieving a high output per hectare, and this can only be achieved by coupling intensive stocking rates with a high level of output per cow. In order to maintain a high output of milk, only sufficient young stock should be reared to meet replacement needs. The question as to the need for rearing young stock on the small farm needs to be fully investigated.

On some farms there may be small marginal areas, which are very hilly and suitable for little else, but normally young stock rearing involves the investment of capital with no immediate return making heavy demands upon cash-flow. Additionally, land put to rearing stock could alternatively be used to maintain a cow in milk production which would certainly contribute to the farm profits.

Small farms are run as family farms or with usually no more than one hired man. Rents are usually high and the farms tend to be equipped with buildings, silage barns, feed fences, which all add up to a very high level of fixed costs.

It is therefore essential that there must be a high output per hectare to cover the fixed costs. Small farms have to be very well managed and there is little scope for error.

2. Medium-Sized Dairy Farms: 41 – 100 Hectares

In contrast to the highly specialised small dairy farms previously described, medium-sized farms run dairy herds not usually as the hub of the farming system but rather as a component of a mixed farming system.

These farms are large enough to run other livestock enterprises for meat production (beef or mutton) in addition to the dairy herd. They are also able to devote arable land to cash crops, whereas the small farmer cannot afford to grow cash crops at the expense of his fodder supplies. Even so, on these medium-sized farms there is a trend towards specialisation and fewer, larger enterprises. In this more streamlined situation there is scope for the development of a high level of enterprise efficiency which brings together the aspects of some economy of scale coupled with the employment of a profit-orientated professional work-force.

A greater availability of land is likely to make available areas where young stock rearing could be carried on, so self-contained herds can be maintained. If this is run in conjunction with a profitable dairy unit, the sale of surplus quality stock becomes an interesting possibility.

The size of the dairy herd on farms in this medium-sized group is governed initially by the availability of grazing land for intensive management. This would be reconciled with the arable area so that a balanced farm rotation could be worked out. Over many years, high-level dairy farming has shown better and more consistent profits to the farmer than any other farm enterprise. Therefore, in most situations the dairy unit has progressively become the more important enterprise even on the medium-sized mixed farm.

3. Large Dairy Farms: 101 Hectares and Over

Extensive farming, assessed in terms of land availability, has long been associated with the corn-growing areas of southern and eastern England. The description 'sheep and barley' land gives an insight into the type of farming systems practised. Light-land farms, where the soil was not inherently fertile had a continuing need for livestock to put fertility and structure back into the soil. Profits from sheep declined rapidly over the years, and they were replaced by dairying in many areas such as the Wiltshire downs where extensive dairying ousted cereal growing as well. An important part of this development was the introduction of the milking bail.

Large farms offered considerable scope for a widely diversified system of farming including cereals, grass seed production, milk, beef or mutton. However, in more recent years there has been a significant trend to simplification, involving in many cases just cereals and dairying, the latter being in convenient viable units of about one hundred cows. Stocking rates tend not to be too intensive, round about 1.8–2.0 cows per hectare.

Many of the herds had the advantage of being started from a green field site, the buildings that were erected being simple and very functional. In many cases it was possible to set up a herd based on the 'block calving' system, which eased management and allowed the units to be designed for a one-man work-force with appropriate relief. The feeding emphasis was on growing and utilising as much grass as possible. Due to the vulnerability of downland areas to dry summers there has tended to be a move towards an autumn-calving pattern. Late lactation cows are not hit as hard by a shortage of summer grass.

THE SYSTEM MUST SUIT THE RESOURCES

From the brief description of the dairying systems given, it is apparent that there is no standard pattern for a dairy farm.

Between the extremes of intensive dairying on typically small or family farms and extensive dairying with greater emphasis on output per man, there is every opportunity to make use of the locally occurring resources which are available; the main requirement is the skill of the dairy farmer to achieve the best blend of resources to match the local conditions and to achieve the highest possible profit level.

Breed of Cattle

Having considered the main physical and environmental aspects which determine the size and location of the farm, it is then necessary to determine the most suitable breed which would best achieve the farm's objectives.

In past years this would have been a topic which would have created considerable discussion. Nowadays though, the most popular dairy breed, by far, is the Friesian/Holstein, accounting for 95.3 per cent of all artificial inseminations in England and Wales.

The main reasons for the increasing popularity of the Friesian/Holstein type of cow is the potential of high yield per cow, coupled with the considerable improvements in milk quality over recent years. A farm's labour requirement is closely related to the number of cows in the herd. High yielding animals therefore allow the farm to produce the same quantity of milk from fewer cows, thereby saving labour. The Friesian/Holstein type also has some merit in producing a very desirable calf when mated to any of the pure beef breeds—Hereford, Angus, Charolais, Limousin or Simmental.

Even within breeds there is considerable discussion as to which strain is to be preferred. Current debate is concerned with the question of whether Friesian and Holstein cattle should have separate or joint herd societies. This is a matter which is of considerable interest to the pedigree breeder.

Once the breed has been decided, other factors to consider are:

Stocking Rate

Stocking rate depends upon a consideration of two basic farm features: (i) soil fertility, and (ii) size of cow.

The area of land required to provide for the maintenance of a cow and for growing bulk food to provide part of the production ration is closely related to the features mentioned.

Stocking rates associated with the dairy breeds are given in Table 4.1.

Table 4.1 Average Stocking Rates

Breed	*Average Liveweight (kg)*	*Cows per hectare*
Jersey	356	3.33
Guernsey Ayrshire	457	2.5–3.0
Friesian Holstein	559 660	2–2.5

The level of fertiliser application also affects stocking density. Nitrogen application on intensive dairy farms has gradually crept up to the 250–350 kg N/hectare level, associated with the appropriate levels of phosphorous and potash, resulting in increased grass growth and potentially higher stocking rates. It is probable however, that during the early 1990s EC regulations will be implemented to control levels of nitrate pollution to water. This restriction, together with control over the spreading of dairy waste products i.e. farm yard manure and slurry, could seriously reduce the productivity of land utilised by dairy farms.

One drawback with intensive stocking rates is that they tend to encourage internal parasites such as worms. Modern routine animal health treatments tend to control this problem however.

Poaching of pastures by intensive stocking in wet seasons can also be a problem, particularly in the spring and autumn. Some form of zero grazing at these times is beneficial, providing the buildings allow such practices and the appropriate equipment is available.

In recent years—certainly since quotas were introduced in 1984—dairy farmers have been forced to think of stocking rates in relation to the attainment of a greater margin over purchased feed per litre. This is vital when the total herd production is restricted.

e.g. Herd Milk Quota		Margin/litre	Total Herd Margin
500,000 l	@	15p	£75,000
		13.5p	£67,500

The difference between a 15p/litre and a 13.5p/litre margin

could be attributed to the increased use of good quality silage, reducing the need for bought in concentrates, giving the higher margin.

The higher herd margin might be achieved by the use of 10 tonnes silage per cow over the winter. Production of this quantity of silage, plus summer grazing might require a stocking rate of 2.2 cows per hectare.

The lower herd margin might be associated with cows being given 8 tonnes of silage per cow over the winter. A tighter stocking rate of 2.4 cows per hectare could be achieved like this, while still providing sufficient summer grazing.

The possibility of obtaining a higher margin per litre basically depends upon the availability of land. Acquiring more land costs money; costs of this nature would certainly equalise the Total Herd Margins mentioned.

Quality of Grassland

Grass growth is very closely related to the soil fertility of the farm. A low fertility status can, however, be improved by use of appropriate fertiliser. Equally important is the type of grassland being used.

There is a whole subject of knowledge which goes under the title of grassland husbandry. If steps are taken to achieve a good fertility status in the soil, then it is equally important to establish grass swards which contain well bred strains of grasses and clovers which can take advantage of the fertility.

Different seeds mixtures are available which will produce leys to fit every possible requirement concerning duration of ley and possible utilisation whether grazing or cutting. Good permanent pasture grows steadily during the grass season and can stand drought conditions, but lacks the ability to respond to fertilisers and produce large amounts of quality grass for grazing and conservation. The stocking rate on a farm can be intensified and the quality of grassland improved by good management.

Grassland husbandry is therefore an important part of dairy management, and provides the basis for the production of large amounts of profitable milk.

EFFECT OF MILK QUOTA ON DAIRY FARM PLAN

Arrangements were put into operation from 2 April 1984 for the scheme to be implemented in the UK by MAFF, although the details concerning the operation of the scheme were by no means complete.

The initial provisional annual quota for each producer for 1984–85 was arrived at by taking the 1983 sales minus 9 per cent. This step was designed to reduce the total output in the EC back to the national quota level as defined by the EC. This level of reduction provided for a small reserve quota for dealing with special cases.

After four years of quotas the surpluses were virtually eliminated. The basis for the organisation of any dairy farm must therefore be the allocation of basic quota litres, appreciating that additional can be acquired by leasing or purchase.

Planning to meet quota is an important feature of management. For example with a quota of 550,000 litres, a 100 cow herd would need to produce 5,500 litres per animal per year. The herd would on average contain 20 per cent heifers so the yield would be allocated as follows:

80 cows	@ 5,600 litres	=	448,000
20 heifers	@ 5,100 litres	=	102,000
			550,000

With the appreciation of the existence of lactation curves, i.e. probable levels of yield throughout the cow's lactation, it is possible to forecast expected peak yields. For cows and heifers to obtain the necessary lactation yields, shown in example, cows would need to peak between 26 and 28 litres per head per day and heifers between 19 and 21 litres. Management of the herd would then have to be organised to achieve the necessary levels of production, for cows and heifers, throughout the year.

From year to year the availability and quality of bulk food, (i.e. grass and conserved grass) is likely to show considerable variation. To fill the farm quota and to provide the necessary economic balance between farm-grown and purchased feeds requires a high level of management skills.

The shock of the imposition of the quota system had a great effect upon dairy farmers. The first thoughts were of cutting back in expenditure and limiting production. However as the year passed the spirit of progress and interest in new techniques become more evident. There was much leasing and selling of quota, coupled with an increase in milk price during the late eighties. These features combined to make the industry much more profitable than in previous years, and dispelled most of the gloom of the immediate post quota introduction years.

PROGRESS TO SELF-SUFFICIENCY

The most important occurrence of recent years has been a very determined move by the majority of dairy farmers towards self-sufficiency in the provision of food for their dairy herds.

Although the dairy industry had accepted for many years that milk production needed to be economic at all levels, it took a 'shock' such as the quota system to bring the point home to many farmers.

The limitation on production emphasised the necessity for every litre to be produced economically. This resulted in a greater appreciation of the term 'Margin per Litre' defined as 'sale price per litre minus cost of purchased feed per litre'.

Achievement of the best margin per litre depends initially upon growing adequate amounts of good grass for grazing and conservation. The second area for continuing thought and development is the use of grass and conserved grass products to produce milk cheaply by reducing purchased feeds.

In some areas, soil type and climate may be less favourable for producing farm produced bulk. In such cases purchased concentrates and bulk food may have to be used at higher rates than normal, irrespective of the cost.

Table 4.2 shows the trend in milk production and purchased concentrates during the period 1982–88.

The trend to greater efficiency in self-sufficiency resulted in a 23 per cent reduction of concentrate usage per litre over the six years period.

Table 4.2 Relationship between milk production and concentrate use, 1982–88

Year	*1982*	*1984*	*1986*	*1988*
Herd size (cows)	111	119	114	117
Milk yield per animal (litres)	5282	5233	5365	5367
Concentrate use kg/cow	1779	1887	1557	1476
kg/litre	0.34	0.36	0.29	0.28
Stocking rates cows/hectare	1.98	2.05	2.00	1.94

Source: FMS Information Unit, MMB.

Summary

The blending of all the farm resources must be the basis for any dairy farm organisation. The resources must be given a priority rating to make sure that the dairy plan has a good practical and economic base.

Current standards should be used to calculate details for any plan. It must be realised, however, that, as shown in Table 4.2, nothing remains the same for long, since there is an ever present trend for current standards of performance to be exceeded and new levels to be set.

A successful dairy farm plan will achieve the optimal level of herd performance which will produce the best on going financial return over a period of many years.

CHAPTER 5

CROPPING FOR THE DAIRY FARM

For many years, the cropping of dairy farms has been planned to achieve a much higher degree of self-sufficiency following the progressively rising prices of imported feeds.

There has been a continuing emphasis on the need to produce from our own farm resources by successive governments. This policy has been closely associated with modern grassland techniques involving heavy nitrogen dressing, various forms of intensive grazing to ensure efficient grassland utilisation, coupled with more efficient grassland conservation systems for winter fodder.

The overall tendency has been to an intensification of land use, indicated by larger herd sizes and more intensive stocking rates.

PLANNING THE CROPPING

The first consideration when planning the cropping for the Dairy farm. must be to decide upon the size of the herd and followers.

In the early eighties dairying underwent a considerable change following the introduction of quotas. The initial reaction was to maintain herd numbers, but reduce the yield per cow in order to achieve quota. This was coupled with a policy of producing more feed from the farm, and thereby increasing the margin per cow. As a result of this there developed a much greater interest in more efficient grazing and conservation techniques of grassland use.

In many of the western parts of the country, the level of nitrogen use has increased to 250–300 kg/N per hectare. This level of application has the ability to produce adequate supplies of bulk food related to grass growth. It is advisable, however, to consider that there is the likelihood of future EC legislation curtailing the use of nitrate fertilisers. Care of the environment is a growing subject, and both fertiliser and farmyard manure are likely to be subject to restrictive control. This being the case the situation will shortly develop that as fertiliser on grass is limited, so alternative crops for the dairy farm will have to make up any shortfall in grass production.

During the late eighties, many dairy farmers became aware of the possibilities of leasing or buying quota, allowing greater milk production from the existing herd and farm.

A tendency to larger herd size is indicated in Table 1.4, implying greater intensity of land use and, once again, the desirability of supplementing grass by other crops.

The factors which affect the growing of crops other than grass are relatively long term. The incidence of several consecutive dry summers has also aroused interest in diversified crops. This will be discussed later in this chapter.

Once the herd size has been equated to the required milk production, the next step is to decide on the young stock policy. The profitability of young stock rearing has been poor for many years when compared to the margins possible from milking cows. The number of replacement stock should, therefore, be kept to a minimum, leaving more land for milk production. On average the gross margin for a heifer is only 45 per cent of that for an average dairy cow. Contract rearing of heifers helps in this aim, and in many cases all replacements are brought in. The usual replacement rate on dairy herds is between 20 and 25 per cent. The lower level of this range must be aimed for, but in budgeting terms a 25 per cent level would be reasonable (so that a herd of one hundred cows would require twenty-five heifer replacements each year). Using the normally accepted livestock equivalents, this would mean approximately one hectare devoted to followers for every three to four cows. Where feeding for the replacement heifers includes straw and purchased supplements this ratio can be increased. With an average calving age of two and a half years, a 'replacement unit' of one calf, one yearling and a heifer would equal about 1.4 livestock units.

Once a decision has been reached on the total number of cows and followers needed, the cropping must be planned, taking into account the following considerations:

Situation of Farm

Choose crops which fit in with the soil and climatic conditions. Generally speaking the western parts of Britain are more favourable to the growth of grass and root crops; the eastern areas to cereal crops. High elevation tends to restrict cultivations: upland areas have a shallow depth of soil and a climate which, while allowing crops to grow, does not necessarily allow harvesting. Late spring with early winter gives a very short growing season and restricts choice of crops. Heavy rainfall on high land requires good drainage, which is not possible where the soil has little depth and overlies rocks.

Size of Farm

The smaller the farm the greater the need for raising output per hectare of land to the maximum. This is where soil type, fertility and skill in cultivation play an important part. Where the land is well drained and free working, the use of short-term leys coupled with catch cropping (eg. kale or stubble turnips) is possible and such a policy would intensify land use and help considerably towards self-sufficiency.

Capital Resources

The amount of working capital required for the purchase of seeds and fertiliser and the capital required to finance the appropriate machinery will govern the range of crops grown. The rise in machinery costs in particular, both capital and running costs, have escalated to high levels over recent years.

Many small dairy farms tend to be over-capitalised in machinery and it is very likely that in many cases contract services could be used. Objections to contractors tend to be concerned with availability and timeliness of operation. In the majority of cases these problems can be overcome provided contracts are placed early enough and that prompt payment for work is guaranteed.

With interest on borrowed money having risen to 15–20 per cent, the arguments for a small farmer to own all his own machines have become less convincing.

Labour

Apart from the labour requirements of individual crops there is need in planning to choose a sequence of crops which give an even spread of labour during the growing season.

A small area of a high-labour demanding crop (e.g. roots) may be justified if it offers employment at otherwise relatively slack periods. Where labour is employed solely on the dairy unit it is possible for one man to cope with the everyday tasks—milking, feeding, slurry disposal—of running a 100-cow unit.

This leaves little time for the equally important aspects of cow management and time off. A reasonable figure of one man per sixty cows is more acceptable; as units become larger so the task or marshalling groups of cows, breeding management and group feeding take up a greater proportion of time. If the unit is to be profitable, these tasks must not be skimped. Where the dairy unit is on a large farm, the provision of grass, grass conservation and supply of fodder crops tends to be the responsibility of the arable staff. It is important that the arable team appreciates the importance of producing good-quality fodder on the farm and how important their efforts are in securing these quality end-products.

Economic Balance of Cropping

The final criterion to be considered when planning any system of cropping is to decide which crops provide the cheapest source of nutrients for the dairy cow and those which fit in best with the local environment.

Dairy cows are ruminants and their diet should be largely made up of cellulose-type foods which are most suited to the digestive system of the cow. Unfortunately, farm grown cellulose-type foods are bulky, with high levels of fibre and cannot always be considered as energy intensive. Grass itself passes through many stages of growth and this allows it to be used as a near concentrate in the spring, or as a bulk, 'belly-fill' food with low energy and high fibre in the late summer.

Modern financial pressures tend to dictate that high margins have to be attained and maintained. This pressure will have much influence on future cropping policies and feeding systems.

In the future it is possible that land prices will increase again, in which case the existence of extensive land use systems will be in doubt and more intensive systems will be required. High levels of production geared to higher margins lead to intensive stocking rates. This releases more land for other uses, such as additional arable or livestock enterprises.

An example of this development has been the introduction of the bunker feed silage system for dairy cows. Stocking rates have been intensified to four cows per hectare, which allows very high margins over purchased feed per hectare, and on one farm in particular allowed the winter wheat acreage to be increased.

The economic balance of production is not a simple comparison of food cost per megajoule of energy, but something which is to be integrated into the whole farm policy and, above all, needs to be flexible to meet future economic trends.

SIMPLIFYING PRODUCTION

The trend in farm organisation has been towards simplifying production to those enterprises best suited to the farm and ensuring that each enterprise is appropriately capitalised. On small farms (40 hectares or less) in high-rainfall areas, this trend has taken the form of increased specialisation on grass.

Better management and fertiliser usage have increased stock-carrying capacity, leading to larger dairy herds which require more accommodation.

To capitalise such expansion in milk production has often meant cutting out other small livestock enterprises and saving on arable equipment by not growing cereals. This has led to a shortage of straw for litter, but the use of sawdust and shavings, cubicles and slatted floors offer solutions to the litter problem.

Similarly, on larger farms, herds have increased in size to become economically viable, but the choice of foods to grow for the dairy herd is much wider.

In lower-rainfall areas, cereal growing, kale and sugar beet offer cheap sources of winter feed without the problem of litter becoming acute. Such diversity of cropping allows the steady employment throughout the year of a constant labour force.

Thoughts on Organisation of Cropping

In the arable situation grass plays a much less important role in feeding the cow. There has been an interest in the system of keeping cows housed all the year round, without any grass grazed in situ. Livestock production on the 'Beef-Lots' of America demonstrate this possibility. Large herds, over three hundred cows, would possibly justify the use of mechanical feeding by spreading the cost of mechanisation. Certainly the grazing of large cow herds is much affected by the problems of poaching in wet weather. This wastes grass and hampers grass regrowth by ruining soil texture. During the seventies, several large herds were set up but all seem to have run into the practical difficulties associated with large numbers.

Management of these large herds has not been tight enough, and with machinery costs escalating, the systems rapidly become economically suspect. Large herds were put on to many arable farms so that they could utilise the grass break which had been put in to promote soil structure and fertility on intensive cereal blocks. In future, large herds on arable farms may be fed on processed straw. It is not certain that grass will play such an important part on these farms as it has done in the past.

On the western side of the country, with a 'milk from grass' policy, the size of herd as a grazing unit will tend to limit the optimum herd size to between sixty and ninety cows. Where herds have increased to between 120 and 150 cows it is desirable to split the herd into two, depending upon the calving pattern. The dry cows could be split off and could be grazed behind the milk cows on a paddock system. In-calf heifers could also run with the dry cow group. Block calving herds will perhaps be grazed as one herd, but the grazing system is more likely to be 'set-stocking', possibly less intensive and more able to cope with the poaching problem.

The price of concentrates has risen considerably during recent years, e.g. from £36 to £160 per tonne in fifteen years. This has had the effect of focusing efforts on the following:

- more efficient production of grass for both grazing and conservation as silage or hay;
- possible extension of the grazing season autumn and spring, for the appropriate groups of cattle.

CROPS SUITED TO THE DAIRY HERD

The previous sections have emphasised the balance which must be reached to cater for the needs of a dairy herd with respect to different resources available on a wide range of farms through a variety of climatic and geographic situations. No general solution exists for this matching of stock, costs and farm since each farm is a unique case, requiring its own specific solution. Grass, grazed and conserved, makes up the major part of the diet for cows in the UK. A variety of arable and forage crops can be grown to supplement grass. Such crops need to be reviewed to present their potential value for use in the dairy herd feeding programme.

Raising crop yields per hectare should be the aim of all farmers provided that it is economic to do so. Heavier applications of fertilisers and a more comprehensive knowledge of their value and use would have this effect on most farms.

The first essential is a knowledge of the existing soil fertility status. A soil analysis survey can be undertaken by various organisations which can then be translated into action to meet any deficiencies that have been identified. Good and carefully timed cultivations contribute their part; full advantage should be taken of weather conditions. An attempt must be made to work with nature, for instance ploughing heavy land in early winter to achieve a frost tilth which will provide a good seed bed. A firm but well-worked seed bed, providing adequate aeration of the soil, is to be coupled with the fineness of seed bed which must be related to the size of seed going in. A close contact between seed and soil particles is desirable. Good arable crops require that sufficient attention is paid to liming and draining.

CEREAL CROPS

Barley

There was a considerable increase in barley area in the mid-fifties. At that time intensive beef systems were developed using large amounts of barley, and because of its higher energy value, it—to a great extent—completely replaced oats as a source of energy for the ruminant.

This trend was greatly accelerated by the work of plant breeders who paid considerable attention to the characteristics of yield, strength of straw and ability to respond to higher levels of fertiliser application. Additional factors, such as better field equipment including more efficient combines and the more widespread use of chemical sprays, all helped to promote the increase in barley area.

Cereal growing affords an example of the economies to be gained from an increasing scale of operations. In the specialist cereal areas there was a trend to continuous cropping with cereals. But such extreme specialisation is not necessary on a dairy farm where a grass ley can constitute a profitable break to cereal growing.

Barley offers the dairy farmer an opportunity of providing a high-energy food in terms of grain and additionally an amount of straw which can be used not only for litter but also for feeding. Recently developments in the processing of straw by alkali treatment have opened up considerable scope for use of treated straw as food. Another possibility is harvesting and processing as 'whole crop'.

When analysed under the gross margin system, barley is shown to have a similar gross margin to many livestock enterprises (excluding dairy cows) and this, coupled with the lower level of fixed costs required, again helped to encourage the expansion of the barley area. In recent years winter barley has become important at the expense of spring barley. (See Table 5.1.)

Table 5.1 Gross Margin Comparison–Cereals v Livestock Enterprises

Enterprise			*Low*	*Average*	*High*
CEREALS					
Barley	Winter	Yield t/ha	4.25	5.65	6.75
		GM £/ha	210	345	450
	Spring	Yield t/ha	3.5	4.7	5.5
		GM £/ha	185	300	375
Wheat	Winter	Yield t/ha	5.0	6.75	8.0
		GM £/ha	260	430	550
	Spring	Yield t/ha	3.35	4.50	5.35
		GM £/ha	330	440	525
LIVESTOCK					
Beef–18-month-old				£	£
		GM/head	—	200	238
		GM/forage ha	—	665	950
–24-month-old					
Autumn born		GM/head	—	233	—
		GM/forage ha	—	358	—
Spring born		GM/head	—	265	—
		GM/forage ha	—	482	—
Sheep—Lowland Spring Lambing					
Stocking Rates		Ewes with lambs per forage ha	8	10.5	13.0
GM/forage ha (£)			150	276	449
Dairy Followers (Friesians)					
GM per animal (£)			261	275	292
GM/forage ha (£)			275	380	508

Source: Farm Management Pocketbook, J. Nix, Wye College 1989.

With barley grain and processed straw being so readily available on the large farms, it is possible that grass will play a less important part on dairy/arable farms in the future than in the past. Dairying has in recent years been the most profitable way of cashing the grass break but this may not always be so.

There have been developments in the processing of feeding barley and straw to make them more useful in livestock feeding.

1. Moist barley storage became popular during the early seventies. This can be taken direct from the combine with a moisture content of between 18 and 24 per cent; after rolling, it can be used for feed, with none of the drying costs associated with conventional grain storage at 14 per cent moisture. The rising cost of the towers needed for this type of storage, however, has greatly restricted interest in this method.
2. The processing of straw, i.e. treatment of straw by caustic soda, seems to offer considerable scope for the future. By treatment of this type the digestibility of straw is improved by up to 75 per cent, making the straw comparable to average-good hay although obviously lacking in protein. The process is carried out on both an industrial and a farm scale. Processed straw could play an important part in feeding dairy cows in the future.

In the successful growing of barley for feed, yield is particularly important. The main features to be observed are, firstly, early ploughing: drill into a frost mould if possible, at a depth of 3.8–5.1 cm, as soon as soil conditions permit in the spring. Secondly, the correct soil acidity should be maintained by adequate liming. Finally, correct quantities of fertiliser and seed should be used; up to 80 units N, 40 units P_2O_5, 40 units K_2O, and a seed rate of 125–157 kg/hectare should be sufficient on good tilths.

Suitable varieties of barley are recommended annually by the National Institute of Agricultural Botany (NIAB) at Cambridge. Leaflets are available on request.

Oats

There has been a great reduction in the amount of oats grown since the mid-sixties, almost completely due to the rising interest in barley. Oats have a lower energy content than barley, and most oat varieties had long straw which was not suited to the expected work rate of the modern combine harvester. Plant breeders gave much attention to the selection of yield and short straw strength, but there seemed to be more potential in barley than in oats. However, heavy feeding of barley (e.g. 12 lbs per day) can result in stomach upsets. Oats therefore offer a very useful alternative. The higher level of fibre in oats takes on a new importance when cows are subjected to high-energy diets.

Oat straw traditionally was thought to have a better feeding value than other cereal straws. This was really explained by the fact that oats, grown in the western half of the United Kingdom, were usually harvested early, before all the nutrients were translocated from the straw up to the ear.

Oats are able to tolerate a higher degree of soil acidity than barley. Continuous cropping, however, poses the risk of cereal root eel-worm infection.

Suitable varieties of oats are recommended for general use by NIAB.

Wheat

When grown on the dairy farm, wheat is considered to be a very good means of cashing the stored up fertility present in grassland. The straw is generally used for bedding. Winter wheat varieties are harvested quite early in the season and provide a suitable entry into the rotation for a fodder catch crop and the autumn establishment of a new ley.

With the heavier yield and better price, wheat has the highest gross margins of the cereal crops.

Lists of recently improved and recommended varieties are available from NIAB, Cambridge.

FORAGE AND OTHER CROPS

Peas and Beans

Peas and beans are the most concentrated protein foods that can be grown. Unfortunately, over the years the general lack of yield resulted in little interest being shown in this type of crop. Traditionally, a mixture of pulse crops were grown with cereals but this practice has virtually ceased. Straight cereals can be supplemented with protein as required. It is also easier to find alternative uses for a crop if it is grown singly, and possibly making the disposal of surpluses simpler.

Root Crops for Cows *(Mangold Turnips, Swedes, Fodder Beet)*

The area of root crops grown for cows has considerably declined, largely due to the high labour requirement for the crop. Mechanisation of precision drilling, pre-emergence sprays and harvesting now allow the crop to be considered for use on a dairy farm. The husbandry requirements are a good seed bed, early sowing, together with a reasonable level of manure and fertiliser application. For example, fodder beet and mangolds should get as a minimum 24 tonnes of dung per hectare and/or up to 600 kg/hectare of fertiliser containing 10 per cent N, 25 per cent P_2O_5, 15 per cent K_2O.

Roots have a liking for humus and whatever dung or slurry can be applied before ploughing will always be well utilised.

Choose high dry-matter varieties of fodder beet and mangolds; sow before May if soil and weather conditions permit; single when the plant is small. With fodder beet use easily lifted varieties; mechanical harvesters are now widely available.

Turnips and swedes are susceptible to attack by turnip flea beetle; control is possible by seed dressing or low-volume sprays. Preventive methods should be applied when the seedlings are about to emerge (i.e. seven to ten days after sowing).

Turnips and swedes are more generally grown in the west and northwest of England, Scotland and Northern Ireland, rather than the lower rainfall areas of the east and southeast. If the crop is to be considered, it should only be grown with the full use of precision drilling and pre-emergence spraying for weed control.

Because turnips are particularly susceptible to frost the dairy farmer should grow only as much as can be consumed on, or soon after, lifting. Mangolds are not frost resistant and are generally lifted and stored in the autumn for use after Christmas. The leaves can be left on if haulage distance is short but otherwise are best removed as they have no feeding value. Mangolds should be stored in a building with suitable precautions against frost, such as insulation with straw bales. Digestive and scouring troubles arise if mangolds are fed before being allowed to mature in the clamp. Swedes can be left longer in the ground, having considerable powers of resistance to frost. They are usually fed from the New Year onward after the turnips have been used up.

Fodder beet with a higher dry matter is increasing in popularity. and seems to be creating interest currently as a farm-produced source of energy. Sugar beet tops are a most useful autumn supplement to the grazing, but ideally should be fed clean and wilted. They should be introduced slowly into the diet and not fed beyond a level of 3.6–4.5 kg per 100 kg live weight.

Spreading the tops on a clean grass field and folding off behind an electric fence is a good way of controlling consumption and at the same time avoiding the digestive troubles so often prevalent when soiled tops are consumed ad lib. An additional safeguard against this trouble is to ensure that the tops are wilted; if they are to be fed fresh then they should be sprinkled with chalk at the rate of 112–168 g per 100 kg.

Arable Silage

Arable silage is a useful home-grown fodder crop. Traditionally it consisted of a mixture of cereals and pulses; research indicated, however, that heavier yields could be obtained from pure cereal crops. The crops can be either autumn- or spring-sown. Both the seed rate and fertiliser dressing should be 150 per cent the rate of a normal cereal crop.

The objective is to cut a heavy yield (35–40 tonnes per hectare) of reasonable quality fodder.

For optimal feed value and bulk, the crop should be cut just before the ear is shot. An arable silage crop will make a good nurse crop for maiden seeds, and shares the heavy overhead cost of direct reseeding. In other situations it could be followed by other catch crops. Arable silage does not have enough feed value for high-yielding dairy cows but can be used as a supplier of adequate bulk to see the winter through; also it is very useful for young stock feeding.

Maize

Maize became a popular crop on the dairy farm during the seventies particularly because of its ability to make good use of slurry as a fertiliser. Maize is susceptible to frost so drilling must be delayed until mid-April or early- to mid-May in the north of England. Maize seed will only germinate when there is a soil temperature of 10°C at 10 cm depth.

Seed rate is 34–39 kg/hectare in 76 cm drills for mechanical harvesting. Apply fertiliser in the seed bed at the rate of 125 kg

N, 60 kg P, 60 kg K per hectare. Protect the crop from birds with black cotton or bird scarers until 23–25 cm tall. Weed control is an important part of the husbandry and use is recommended of pre- and post-emergence sprays (paraquat at 2.8 litres/hectare and atrazine at 2.2–3.4 kg/hectare respectively).

The NIAB Farmers' Leaflet lists the range of forage maize available.

Harvesting will take place in late September–early October to give the maximum yield of good quality silage after a growing period of about 150 days. Yields are of the order of 37–42 tonne per hectare.

Kale and Stubble Turnips

Kale as a crop, lends itself to the increasing use of machinery, both in growing and harvesting the crop. Kale is usually broadcast, but can be drilled at 4.5 kg per hectare in rows 69 cm apart. The crop is very responsive to fertilisers and will give a good return to 600 kg per hectare of 20:10:10 fertiliser. In the seed bed, a further top dressing of 100 kg of nitrogen will ensure a good yield.

These levels can be reduced if a good dressing of dung (50–63 tonne per hectare) is applied. Yields of 20–35 tonne per hectare are possible. Kale can be usefully grazed in autumn and then zero-grazed into mid-winter.

The varieties of kale have differing degrees of resistance to frost. Popular varieties include Canson, Thousand Head, Marrow Stem, Gigantic, Maris Kestrel.

Kale for many years was found difficult to utilise in a wet autumn. With the dry summers of the late eighties, it has been increasingly used as a forage crop grazed in situ during late summer and early autumn (mid August–September). One of the problems of grazing kale is the shorting of the electric fencing wire. Another is that feeding frosted kale to cattle causes digestive upset and causes milk yields to drop alarmingly.

FORAGE FARMING

The name describes the system whereby maximum use is made of forage crops which are grazed in situ. This reduces the amount of handling and hence heavy labour costs. The harvesting of forage crops has now been mechanised, and is commonly described as zero-grazing; it is used extensively in the autumn and spring periods. This overcomes the problems of cattle expending much energy when going out to graze the forage crop.

Forage crops are very suitable for feeding to cows in mid and late lactation and also for dry cows. Cows in early lactation need less forage and the diet must be much higher in energy.

When forage crops are used either grazed or in situ, it is very important to regulate the supply of the crop to the cow. This is particularly important in the spring.

In many cases when a dairy herd is put on to supplementary forage feeding, not enough thought is given to the availability of 'follow-on crops'. Forage crops, therefore, offer a means of changing from full winter rations to spring grazing.

The forage should allow the same level of milk from bulk feeds to be continued; to ask for increased milk from forage at this time is unwise. Winter rye is a leafy succulent crop which has often been over exploited.

Table 5.2 Sequence of Forage Crops

Period of Use	*Crop*	*Date of Sowing*
March–April	Italian Ryegrass Rye	September Requires good establishment for early spring growth
mid-June–August	Stubble Turnips Tyfon	April–May Drilled in 3 sections from mid-April at 2–3 week intervals to give succession of grazing
September–early Oct.	Kale Maize (zero graze)	May

When replacing grass, the technique of spraying turf with paraquat and then direct drilling saves cultivation cost and reduces poaching in winter. These crops will give grazing at times of the year when grass is dormant in the early spring or is making little growth as a result of drought. All respond to generous applications of nitrogen. Most of these crops can be grown as catch crops, thus increasing production over the rotation

Table 5.3 Forage Variable Costs (£ per hectare (acre) per annum)

	Grass-Dairying[1]	*Grass-Other*[1]	*Forage Maize*[2]	*Kale*
Yield tonnes/ha (tons/acre)	—	—	(silage) 35 (14)	50 (20)
Seed	8 (3)	7 (3)	115 (47)	20 (8)
Fertiliser	100 (41)	60 (24)	50 (20)	90 (37)
Sprays	12 (5)	8 (3)	15 (6)	20 (8)
Total	120 (49)	75 (30)	180 (73)	130 (53)

	Rape & Turnips	*Mangels*	*Swedes & Turnips*	*Fodder Beet*	*Cabbage*
Yield tonnes/ha (tons/acre)	—	100 (40)	65 (26)	75 (30)	90 (36)
Seed/Plants	15 (6)	50 (20)	20 (8)	75 (30)	125 (50)
Fertiliser	85 (34)	115 (47)	95 (39)	110 (45)	110 (45)
Sprays	—	45 (18)	25 (10)	100 (40)	35 (14)
Total	100 (40)	210 (85)	140 (57)	285 (115)	270 (109)

1. These are *average* figures only. Intensively grazed grass may have higher fertiliser costs in particular—possibly approaching £150 (60) for dairying. The seed costs will vary according to the proportion, if any, of permanent pasture and the length of the leys. Fertiliser inputs are often less on permanent pasture too, but this depends largely on the attitude of the farmer/manager, which will also be reflected in the stocking rates and productivity levels per animal achieved. Note that the dairying figure is the average for the grass devoted to the *whole herd*, i.e., followers as well as cows; the figure for cows alone will be higher (approx. £135 (55) total forage variable costs).
2. Contract work on maize: drilling £30 (12); havesting (inc. carting and clamping) 100–150 (40–60). For further details on this crop see 'Forage Maize', A. N. Evans, F.M.S. Information Unit Report No. 59, M.M.B., 1988.

Source: Farm Management Pocket Book, J. Nix, Wye College, 1989.

without diminishing the area of land devoted to cash crops—which arable/dairy farmers would be loath to do.

The use of forage crops can undoubtedly help to increase output per hectare on dairy farms and reduce feeding costs. The system allows wide scope for originality and initiative in the cropping rotation.

The costs of growing some of the more popular forage crops are shown in Table 5.3.

The costs of these crops per acre tend to be small relating to the advantage to be gained from feeding them during times of high milk price.

Disposal of dung or slurry has become quite a problem on predominantly grass farms following the increase in dairy herd size. Large amounts could be applied to land prior to ploughing for forage crops, easing this disposal problem. The freedom to spread dung and slurry is, however, certain to be restricted by future EC Regulations.

CHAPTER 6

THE CHEAPEST FOOD FOR MILK

Temporary and permanent pastures occupy 7.0 million hectares compared with an arable area of 5.0 million hectares in Great Britain as a whole. It has been estimated that this area of grass supplies around 45 per cent of the total energy consumed by livestock in this country, though occupying some 60 per cent of the total agricultural land. It follows, therefore, that production per hectare of grass is appreciably lower than production per hectare of arable—a significant fact in the economy of this country at the present time. Yet, to the dairy farmer, grass represents his main and often sole source of nutrients for the dairy cow during at least four months of the year, and is undoubtedly the cheapest source of nutrients consumed by the grazing animal.

Researchers have shown that conserved grass in the form of hay and silage are two of the cheapest sources of starch and protein available for winter feeding. Thus grassland should be able to make an increasing contribution to the economics of dairy farming generally.

INCREASING PRODUCTION FROM GRASS

To achieve this, there are three strategies for dairy farmers to consider.

First, yield of grass per hectare could be increased by better establishment, better management of established leys and permanent grass, and a higher level of fertiliser treatment.

For a discussion on pasture establishment and seed mixture, the reader should refer to books on grassland husbandry, but in general dairy farm management, the concept of *grass as a crop*, entitled to proper fertiliser treatment and to adequate periods of rest following defoliation—grazing or mowing—would help to establish grassland management on a higher level.

Many pastures are seriously reduced in productivity by over-grazing and treading in the winter and by under-grazing in the summer.

Secondly, excess grass should be properly conserved during its peak periods of production, May–June, and again in September, when growth is out of proportion to the stock carried. In recent years with the increasing level of milk yields, autumn-calved cows have been housed at the time of calving to provide more control over the feeding of this class of animal.

This has resulted in much autumn grass being available but not used. Spring-calved animals in late lactation can certainly make good use of this grass but with a mild autumn they cannot utilise all that is available. Apart from lack of utilisation, it is also a fact that when grass goes into the winter with excessive growth—'winter proud'—it becomes more susceptible to 'winter kill' by frost action. A growing management feature has been the introduction of hill sheep to lowland grass farms for perhaps three to four winter months. Advantages from this system include better grass utilisation, improved tillering due to close cropping by sheep, and usually, some financial arrangement.

These peak surpluses are normally conserved in the form of hay, silage or dried grass, and the merits of the each method are considered later.

Thirdly, there should be better utilisation by the grazing animal. This involves heavier stocking and improved grazing techniques to reduce waste in consumption and by limiting the intake of grass to give more efficient digestion of what is eaten. This aspect of rationing the cow's feeding—particularly concerned with grass—is dealt with fully in Chapter 10.

GRASS CONSERVATION—HAY

Over the past 30 years, the feeding of hay to the dairy herd has gone through a most dramatic change. From a position of being the major source of maintenance feed for dairy cows, it has now been relegated to being fed to dairy replacements only during the first few weeks of their lives. Many dairy farms do not make any hay at all, buying in a few tonnes as necessary.

The reasons why hay has fallen into such disfavour are two-fold:

(i) The nutritive value of hay can be very variable according to conditions. At its best, it can be equal to brewers' grains, at its worst, equal to average straw.

Table 6.1 Quality of Hay

Type	*DM g/kg*	*ME MJ/kgDM*	*DCP g/kgDM*
High quality	850	10.3	77
Medium quality	850	8.7	70
Low quality	850	8.1	55

Source: Nutrient Allowances, Composition of Feeding Stuffs, MAFF 1988.
Note: DM, ME and DCP are defined on pages 123–25.

Silage making is often interrupted by rain. If this continues for several weeks then the grass may become too mature for silage, with low energy and protein value and with high fibre. The last resort in such an event may be an attempt to make hay, resulting in a low quality product. It can be seen from the requirements of making up a ration for a high yielding cow that hay has little part to play.

(ii) The costs of hay and silage are shown in Table 6.2.

Table 6.2 Silage and Hay Costs

The average costs of producing and harvesting, on a full costs basis (i.e. including rental value of the land, all labour, share of general overheads, etc.) are approximately as follows:

Hay: £55 per tonne; Silage: £16 per tonne.

The % breakdown of total costs is as follows:

Variable Costs	*Fertiliser*	*Seeds and Sprays*	*Contract*	*Sundries*
Silage	32	1	7	2
Hay	21	1	3	3

Fixed Costs	*Rent*	*Labour*	*Tractors*	*Deprec. & Reps.*	*FYM/Lime*
Silage	26	8	13	6	5
Hay	20	17	21	10	4

Source: B. Roscoe and M. Turner, Milk Production 1985/86, University of Exeter.

A comparison of production costs per tonne between hay and silage is made more meaningful when additional steps are taken to compare costs on a dry matter basis:

Hay @ 85% DM—£55/tonne
Silage @ 25% DM—£16/tonne

Corrected to same 100% DM
Cost per tonne Hay £64.70 per tonne
Silage £64.00 per tonne

Having established that the costs on a dry matter basis are similar, then the undisputed popularity of silage is simply related to its better feed value. Silage volumes are shown in the next section.

For farms, other than intensive dairy farms there is a place for hay. The process of making hay has been considerably improved by the increasing degree of mechanisation, but even this development cannot remove the biggest hazard to hay-making—the weather. Traditionally hay is left to mature in the field. The modern trend is to work the hay intensively after cutting. This, coupled with the appropriate weather, can ensure a good product.

The essential points in conserving are:

- cut early in the season,
- only cut an area that can be easily handled, taking into account the current weather,
- in good weather, work the cut grass to 'make' the hay suitable for baling,
- minimise the time it is left out on the field.

Risks with Baling

For larger farms, the choice of haymaking equipment largely depends on the area to be handled and the capital available.

Desirable as high output per man is, its pursuit in haymaking is often attended by reduced feeding value in the product, as compared with the quality achieved by more laborious hand methods. For example, the great danger in baling is mouldiness in the bales through baling hay which is still damp. Hay for baling must be thoroughly dry right through the swath; tedding is recommended in addition to swath turning. Forage harvesters have been adapted to operate at lower rotor speeds for crushing hay and facilitating drying in the swath. Roller or crimping machines are now available for the same purpose.

In favourable weather hay so treated can be baled within forty-eight hours of cutting. Do not bale until the dew has completely lifted in the morning and before it comes down in the evening. Give the bales time to 'sweat' in the field; small bales should be set up in the field if rain threatens. Barn hay-drying does offer some solution to risk of exposure to weather.

The weather hazards associated with haymaking can be minimised by use of the Farmers' Weather Service, available at the Meteorological Office at Bracknell. Notice of dry-weather periods can be obtained.

GRASS CONSERVATION—SILAGE

The processes involved in silage making are more complicated than in haymaking. It is important to understand in outline the changes which occur during the silage making process in the pit.

Plant Respiration

When a crop is cut for silage, some of the plant's life processes still continue. Respiration is the name given to the process whereby a plant takes in oxygen and breathes or respires, producing carbon dioxide and heat. The plant nutrients which are mainly sugars are broken down. The significance here is that where there is plenty of oxygen present in a silage pit, the plant nutrients are lost and excessive heat is produced making the remaining plant nutrients less digestible. This explains the importance of rolling a silage pit to exclude air. Attempts should be made to keep the temperature up to the range of 32–42°C and no higher.

Fermentation

Silage making is a very similar process to pickling onions in vinegar. In the case of silage, the acid is produced by the action of bacteria already present in the harvested grass crop or indeed any other plant material. (The same principles apply to maize and kale silage.)

The fermentation process is complicated and many chemical compounds are formed. The most important of these are three acids—Lactic acid, Acetic acid and Butyric acid—a good silage will have a high lactic acid content, varying levels of acetic acid and no butyric. A badly made silage will have a high butyric acid level and a low lactic acid content.

The way in which the quality of preservation of silage is assessed is on the basis of the acidity. This is measured by a figure known as pH. The total range of acidity extends from 0 to 14; good silage will be best preserved by a pH of about 4.1–4.5. Levels higher than this would allow its decomposition process to commence, levels of pH below 4, make the silage less palatable to stock.

Practical Relationship between Respiration and Fermentation

The production of lactic acid is encouraged by the temperature approaching 32°C, lower temperatures lead to a higher level of butyric acid, resulting in 'cold' or 'sour' silage.

It can be appreciated that the degree of air in a pit exerts a major influence on the fermentation process; lactic acid bacteria need air and some heat whereas butyric acid bacteria need a low temperature and no air.

It follows, therefore, that when grass is left to wilt in the field before ensiling, it is likely to result in better silage than if the grass was ensiled in a very wet condition, since it is less likely to be short of air.

In addition to wilting, the stage of growth at which grass is cut must also be considered, as its association with the presence or absence of air dictates the degree of consolidation required. The amount of rolling on the pit must be related to the temperature inside the pit. A silage thermometer is therefore a very necessary part of the silage maker's equipment.

Provided grassland is managed properly and adequately fertilised, silage making can begin two to four weeks before the normal time for haymaking. This is the first step in making better use of the grass crop.

There has been considerable progress in the mechanisation of the silage-making process. The forage harvester has developed from single to double and eventually to precision chop. The size of trailers has been increased up to 6 tonne capacity. The process of wilting was introduced during the mid-sixties and superseded the previous direct cut system.

Attention was directed towards making high dry matter silages (25–30 per cent DM); this made the grass cutting an operation additional to the actual grass pick-up. Where grass was stored in towers, dry matter of 30 per cent plus was necessary. However, on the average farm with clamp silages, it is recommended that the dry matter is in the 22–25 per cent range.

Silage effluent has become a serious problem, being very polluting to streams and rivers and having a very serious effect on wild life. It should be collected in a sealed tank allowing 23 litres of capacity per tonne of material ensiled. Where herbage dry matter is below 20 per cent the dry matter loss in effluent can be as high as 20 per cent; with herbage dry matter content exceeding 25 per cent, there is no effluent problem.

With clamp silages, the buckrake is used to place the grass in the clamp, after being ferried from the field by the tipping trailers.

The development of self-feed silage eliminates the problem of handling silage a second time out of the clamp. The self-feed system does however provide a constraint to the number of cows which can be fed on the self-feed silage face. In the early seventies as size of herd increased, many farmers moved to 'easy-feed' silage. This involves providing a feed fence where the silage can be placed each day. Such a system allows a check to be made on the amount of silage fed to the cows.

A wide range of equipment is available such as forage boxes, fore-end loaders and block cutters which greatly reduce the labour requirements. However, these systems require more capital and must not be embarked upon lightly.

The weight of silage removed from the clamp is at least 25–33 per cent less than the grass ensiled. This can be attributed to effluent and the respiration losses, which in turn encourage the desirable lactic acid fermentation process. Further losses due to weathering and deterioration of the outside layers of silage are minimised by proper and thorough consolidation and protection with a covering plastic sheet to reduce seepage of rain into the clamp. It is desirable to have silage clamps covered but this involves considerable capital expense, and is difficult to justify.

Silage Making Guidelines

Much effort is put into the provision of good grass for silage making, but this effort can be wasted through poor techniques of filling and sealing. Unfortunately, most of the wastage occurs inside the pit and therefore goes on unseen. This makes it important that the basic rules of good silage making must be

Using a front-end fore-loader for making clamp silage.

Unloading at the clamp from a silage trailer.

Here a blower is being used to fill a self-feed silage clamp with precision chopped material dropped into a moving bed dump box.

A forage box and grain blower filling a bunker silo. The material being sprayed in requires little additional compaction, thus saving labour, improving fermentation, and resulting in a friable product that is easy to extract.

adhered to. The following are the important points to remember in making silage.

1. The wilted grass must be distributed evenly throughout the clamp. Avoid pockets of air and achieve good consolidation. These are the first steps to control the initial temperature which in turn controls the initial period of aerobic fermentation. Temperatures above 32° C lead to the development of moulds and are a direct result of an unrestricted respiration process, which burns off much of the available carbohydrates and renders the protein indigestible.
2. The best way to contain the temperature and prevent excessive fermentation losses is to prevent air getting into the pit. The practical way to achieve this is to cover the pit with plastic sheets every night. This is a discipline which must be operated on every silage pit.
3. Losses in the pit are aggravated by slow filling and by large pits with large surface areas. The most effective way of limiting these features is to use the Dorset Wedge system of filling the pit. The technique is to fill the material into the pit in a wedge shape starting at the back wall of the pit. As the grass comes in so the pit fills and the front of the wedge is maintained as a steep ramp. The grass filled in to the appropriate height is sheeted down progressively as the ramp moves forward. The sheet can be secured by straw bales or old tyres. Final sealing should take place as soon as filling has been completed.
4. Care should be taken to eliminate wastage on the shoulder, i.e. the join between the silage and the pit wall. The walls ideally should be sealed to prevent ingress of air, and plastic sheets used for this purpose can be folded under the top sheet to give an airtight seal.
5. Problems arise with grass coming into the pit which is either too dry or too wet. Excessively wilted grass will tend to overheat in the pit; extra consolidation will have to be provided. Highly wilted grass, of course, suffers from excessive field losses as well. Grass which is put into the pit with too much moisture requires little consolidation. It quickly settles, drives out the air, prevents heating and inevitably results is cold butyric acid silage, a foul-smelling product which is unpalatable to stock and is low in protein. As mentioned previously, a 22–25 per cent dry matter grass is likely to make the best type of silage.

40 million tonnes of silage are fed annually in the UK, an increase of 300 per cent in 15 years, indicating the extent to which silage has replaced hay. This is a big step forward offering the real prospect of reducing winter feeding costs for dairy cows.

Making good silage depends upon having good quality grass to put into the pit coupled with a very high standard of pit management. Silage will never be better than the grass from which it was made.

The need to make good silage in spite of the inconsistency of the weather has led to development of a new industry concerned with silage additives.

Additives in Silage Making

The objective in the silage-making process is to preserve grass at its best nutritive level. This is achieved by encouraging the desirable lactic acid fermentation process to produce a pH of about 4.3–4.5 which will inhibit mould and putrefactive bacteria. The desired level of acidity, 2½ per cent lactic acid concentration, will ensure a state of stability in the silage pit.

Additives have been developed and used to achieve this stability more quickly. Molasses has been used for some time as a means of supplementing plant sugars, since it can be readily converted into lactic acid.

In Finland, the AIV process uses mineral acids to halt growth of all bacteria and moulds, but it is necessary to neutralise these acids before feeding.

There is quite a range of additives available in the United Kingdom. The general aim is to supplement the natural fermentation process either by providing readily fermentable material or supplying acid which will more quickly provide a pH level in the silage to check the fermentation process.

Additives can be classified according to their constituents and can be grouped as followed.

(a) Acid—inorganic and organic acids—either alone or in combinations. The aim is simply to reduce pH by artificial means to a level which will prevent any deterioration in the ensiled crop.

(b) Biological—bacteria and enzymes which are added in various proportions to the silage encouraging the natural fermentation of organic acids from the ensiled material, thereby reducing the pH.

A Simple Comparison of Acid and Biological Additives

Acid Additives	*Biological Additives*
(i) Dangerous – to skin and eyes	(i) Safe to handle
(ii) Corrosive to machinery	(ii) Non corrosive
(iii) Acid reduces pH artificially	(iii) Natural lactic acid fermentation reduces pH
(iv) Inorganic acids have no nutritional value	(iv) Lactic acid converted into glucose and acetic acid in rumen
(v) High volume additive	(v) Low volume additive

During the eighties there have been many developments in the manufacture of silage additives.

The subject has become quite complex—readers are advised to contact ADAS, who have published several detailed reports on the effectiveness of the various additives.

The areas concerned in the reports include (i) preservation, (ii) animal performances, (iii) protein protection, (iv) ease of application and (v) degree of corrosiveness.

A star rating is applied to a wide range of additives, so these reports would be a useful addition to any silage makers bookshelf. In view of the acids used as preservatives it is most important to read the instruction notes as to concentration, levels of usage and provision of protective clothing.

Widespread field trials indicate that additives do help to provide a better-quality silage product which is then capable of providing a better cash return on the dairy herd.

Due to the explosion in additive technology and the resultant bewildering ravage of products available, the EC in 1989 decided to produce a directive to control the production and use of additives.

Silo Construction

When construction of a silo is to be undertaken it is important that the site should be properly investigated before construction work begins, to ensure that it has adequate load bearing capacity.

Floors should be not less than 100 mm thick, but if heavy vehicles are likely to have access they should be 150 mm thick and reinforced. Many large loading shovels used for filling and consolidation impose extremely high axle loads. The grade of concrete must be C7P to reduce acid attack and concrete should be laid on a damp-proof membrane to prevent seepage through the floor. The floor fall should be 1.5 in 100 away from the face.

Retaining walls can consist of proprietary units of either timber or concrete, or can be constructed in situ of mass concrete or concrete blocks. Walls should be adequately reinforced, and new designs or existing buildings should be checked by a Chartered Civil or Structural Engineer. A drain, not more than 500 mm from the inside face of the wall, is necessary to convey away effluent and prevent a build-up of hydrostatic pressure. It is important that all walls should have a sight rail above the wall to warn the tractor driver.

A roof to the silo is by no means necessary and hay or straw may be stacked on top but not too deeply initially.

Earth walls provide a low cost alternative but new regulations are currently about to be introduced which will significantly affect the necessary approval for pits of this type with regard to structural detail and effluent disposal.

Assessment of Silage Quality

Once silage has been made in the pit, the next step would be to have a sample analysed. This sample should be taken about six weeks after making to allow the silage process to be completed. Care should be taken to make sure that a representative sample is taken. This is made more difficult when first, second and even thirds cuts are included in the same pit.

Silage quality can be assessed from a conventional silage analysis.

How to Read a Silage Analysis

The first step is to look at the pH and ammonia levels, together with the dry matter; these results give an instant measure of the success or otherwise of the fermentation.

Ideally the silage should have reacted a degree of fermentation which will preserve the grass in a stable condition while retaining as much feed volume as possible. The feeding value can be assessed by looking at the ME (Metabolisable Energy) and the digestible crude protein levels.

Though the analysis is important, the physical characteristics of the silage such as the colour, smell and keeping qualities should also be taken into account.

(a) Dry Matter Percentage (DM per cent)

This is controlled by the weather and the degree of wilting. Remember that if wilting is allowed to proceed too far then it will eventually produce hay rather than silage.

Grass wilted to over 30 per cent DM is more difficult to consolidate successfully to exclude air from the silo, and is also prone to spoilage by yeasts and moulds. Wetter silages below 25 per cent DM have much better keeping qualities.

(b) Crude Protein Content (CP per cent)

This level is related to the maturity of the crop being ensiled, the levels of nitrogen fertiliser applied and the proportion of legumes in the crop. The higher protein levels will always be found in young leafy grass.

(c) Modified Acid Detergent (MAD Fibre Content per cent)

The level of fibre is determined by the stage of growth of the crop when it is cut and may vary from 22–50 per cent. The lower the fibre content so the higher the energy level in the grass and consequently in the silage.

If grass is cut before heading occurs, then the fibre level tends to be in the range of 26–28 per cent. Average levels of fibre for silage are 30–35 per cent.

The ME and 'D' value of silages are determined by applying mathematical equations to the MAD fibre level.

(d) Ammonia Nitrogen per cent (as a percentage of Total Nitrogen)

This is a very important guide to the quality of silage fermentation. A high level of ammonia makes silage very unpalatable to stock. The ammonia is a result of the break down of protein which occurs if the pH (level of acidity) and clamp temperature are too high 6–7 days after ensiling.

Clamp conditions of pH 5.0 and a temperature of approximately 40°C are ideal conditions for *Clostridia* bacteria to break down protein and produce ammonia and butyric acid. Efficient rolling of the clamp and complete sheeting at night will help to reduce clamp temperature.

Where there has been a long drought spell followed by heavy

Table 6.3 Silage Analysis/Ammonia Nitrogen and Silage Fermentation

Level of Ammonia Nitrogen	*Likely Result*
Less than 5%	Excellent fermentation
5–10%	Very good fermentation
10–15%	Slightly butyric (foul smell)
15–20%	Butyric with protein breakdown
Above 20%	Unsatisfactory fermentation with considerable protein breakdown

rain, the fertiliser applied during the dry weather may have been taken up but not immediately converted into plant protein. This can affect fermentation. In this instance there would be a high ammonia level and a high pH value but little or no butyric acid present.

This is because nitrates from the fertiliser in the forage have been broken down to ammonia in the clamp, which can seriously affect an otherwise normal fermentation. This possibility should be considered when applying fertiliser to grass which is to be made into silage.

(e) Silage pH (acidity)

A satisfactory fermentation is indicated by a pH of 3.7–4.2 for wet silages and 4.3–4.8 for drier silages. Silages untreated by additives can take up to 6–7 weeks before settling at a stable pH. Silages treated with additives are likely to achieve a stable pH within a few days.

The longer it takes a silage to stabilise in pH, so the greater risk of spoilage microorganisms developing and producing a poor fermentation.

(f) Metabolisable Energy (ME)

This is a measure of the energy level in the silage expressed as megajoules per kilogram (MJ/kg) of dry matter. It is calculated by an equation relating to the fibre level in the silage. Silage levels tend to vary between 8.5–11.5 MJ/kg.

(g) 'D' Value—Percentage Digestible Organic Matter in the Dry Matter

The digestibility of grass is closely connected to maturity and storage of growth. The coefficient 'D' is calculated from the MAD fibre content and is another way of measuring the energy potential in a silage. The higher the 'D' value the higher the level of energy. The range is from 55–72 per cent. An average 'D' value is 62–69 per cent.

(h) Digestible Crude Protein (DCP)

This is estimated from the Crude Protein and the ammonia in the silage. Levels in grass vary between 55–160 g/kg. An average DCP is 100 g/kg. The figure is often expressed as a percentage, e.g. 165 g/kg = 16.5 per cent.

Regardless of what the silage analysis says, it is only a chemical assessment of the silage. *The real test of silage quality is the way in which it can be used to reduce the consumption of expensive purchased feeds.*

Quality and Quantity

There is a very close relationship between the yield of grass per cut and the quality of the ensuing silage. For many years some farmers thought of silage almost as a salvage operation, after constant rain has ruined the haymaking schedule. Modern thinking is concerned with attempting to obtain as much good quality silage as possible. Multi-cut silage, a system of taking at least four cuts during the season, was evolved during the mid-seventies. Unfortunately, such a system cannot cope with a succession of dry summers and subsequently lost favour. A more traditional three cuts are taken on most silage farms.

Table 6.4 illustrates the range of feed values possible and emphasises the importance of good grass being made into good silage.

Table 6.4 Average Values of Silage

Silage	*DM*	*MJ/kg*	*DCP per cent*
grass high quality	250	11.8	125
grass medium quality	250	10.6	105
grass low quality	250	9.2	100

Source: MAFF, *Composition and Nutrient Value of Selected Feeds*, P2087, 1988.

Skilled silage makers often attain levels exceeding 11 MJ/kg, which is obviously a highly prized feed level.

The amount of silage required during the year by a dairy cow can vary enormously according to the system of milk production practised, e.g.:

Low Yield system—very extensive, permitted by large area of land available, majority of milk from grass and conserved grass products. Silage required, up to 12 tonnes per cow for winter use. Stocking rate: 1.65 cows per forage hectare.

High Yield system—intensive, limited land availability, regular need to buy in bulk feed. Silage required for winter, about 8–9 tonnes per cow. Stocking rate: 2.8 cows per forage hectare.

In recent years, with the advent of 'Buffer Feeding' through the summer, it is likely that the majority of herds plan to have at least 10 tonnes per cow available for the year.

The greatest danger that every dairy farmer must guard against is the possibility of running short of silage before the winter ends.

In a dry season, if it seems that the required silage tonnage is not going to be available, then steps should be taken by mid-summer to supplement the silage by purchased feeds, e.g. hay, brewers' grains or sugar beet pulp.

In the economy of crop production on dairy farms today, silage making must play an ever-increasing role. Efforts must be made to make as much silage as possible from grass in the spring. Where there is a flush of grass in September, this can be cut for silage, and used satisfactorily for dairy followers. Silage, if adequately sheeted down and protected, can be stored over to the following winter. It would always be a useful supplement in a very dry summer.

Big Bale Silage

This silage making technique has become increasingly popular over recent years. Between 10–12 million bales are made in England and Wales each year. This represents about 15 per cent of the total dry matter made as silage, but this proportion will increase if the current trends continue.

Big bales were originally introduced as a means of offering a silage availability to many northern farms who were still largely dependent upon hay. As an alternative to hay making big bale silage reduces the weather risk and requires little capital outlay for storage facilities.

The system is well suited to:

(i) farms which are without silage making or storage facilities.
(ii) supplementing an existing silo or for ensiling small quantities of grass.

Frequently a couple of fields get out of sequence with the main silage operation, and big bales offer a method of conserving these spare acres.

The original system involved putting large plastic bags around bales of wilted grass. Each bale typically weighed about half a tonne and was, ideally, firm and cylindrical to facilitate bagging and storage. The bagging system although reasonably economical, suffered from the risk of bottle-necks caused by the bagging and tying operations.

Wrapping developed as an alternative to bagging. The close wrapping of the plastic restricts the entry of air into the bale unlike a loose fitting bag.

The cost of the wrapping by film is slightly less than that of the bag, but a general cost of the operation would be about £3.50 per bale including both baling and wrapping. The quality of any bag silage is of course subject to the same problem as with clamp silage.

Baling is independent of the usual silage transport operation. Handling can be done with existing equipment and it is possible to convert existing trailers into big bale transporters. Obviously though, a spikeless method of handling is required once the bales are wrapped. Bales wrapped in a field are prone to damage during handling and transporting but full advantage can be taken of good dry weather by finishing all the baling before transporting the bales. To make the best use of bales, they need to be stored on carefully selected sites, ideally sheltered, with good access and away from any obvious source of rodent activity.

Big bales bring a new availability of flexibility in silage feeding which was not possible before. Ring feeders are an obvious means of presenting the bales to the stock. They offer the possibility of easy-feed, e.g. heifers on a self-feed silage with an electric fence while they are becoming more accustomed to the self-feed system. Big bales tend to be made from rather more mature crops than normal first and second cut clamp silages. This has the potential to increase milk quality due to higher fibre intake.

The technique of later cutting is now changing. Early cut big bales made of first cut silage can be used in drought periods to offer a high quality supplement to grass for newly calved cows, taking advantage of the high seasonality payments available from July–October.

Using baled silage in this way would reduce the necessity of breaking into the traditional first and second cut silo pits. The use of big bales is therefore very flexible and fits into many feed situations throughout the year.

Ensiling Brewers' Grains

If, as has been previously suggested, the silage availability is less than that required, a very suitable supplement can be provided in brewers' grains. Deliveries can be arranged on a regular basis throughout the winter, but a better way to secure the supply is to have them delivered during the summer months, and to ensile them. The grains can be put under grass silage, in which case they must be delivered before the first grass cut is made. Alternatively, grains can be stored in any form of airtight silo, be it concrete or the temporary wire-mesh cylindrical silo.

One tonne of wet grains will occupy 1.9 m^3. The silo should be well consolidated and sealed immediately after filling. Grains can be stored successfully in a variety of buildings, such as Dutch barns lined with sleepers. Brewers' grains exert considerable lateral pressure, depth for depth greater than grass silage, and can be dangerous. Advice on new buildings and conversion of existing buildings can be obtained from the local ADAS adviser.

Brewers' grains have a feeding value equivalent to good silage on a DM basis. (DM–28 per cent; ME–10.0 MJ/kg; DCP–14.9 per cent.) (Freshly delivered grains have a 22 per cent DM.)

Grains are very consistent in quality and are often used to mix in with silages of poor palatability to make them more acceptable.

Maize Silage

Modern forage harvesters fitted with the suitable maize attachment can handle the average maize crops quite successfully. Maize silage is easy to make provided it is well chopped (preferably precision chopped), the silo filled rapidly, and sealed as soon as possible after filling. Maize silage is highly palatable but of a low protein content—typical analysis would be DM–21 per cent, ME–10.8 MJ/kg, DCP–7 per cent with D value of 65. The digestibility of the maize crop is high and changes little during the later stages of growth. It is a high-energy type of silage which, if used for milking cows, needs supplementing with protein. Experience suggests that it is best fed in conjunction with grass silage up to a 60 per cent grass–40 per cent maize silage ratio. Maize silage is very suitable for feeding to beef cattle and young stock, but again, a protein supplement is necessary.

GRASS DRYING

As a conservation process, grass drying is the one accompanied by least loss between the field and the cow and, as with silage making, is to a large extent free of weather risks.

When grass is artificially dried, it can be cut and conserved in the very early stages of growth when it is 10 to 15 cm long and at its highest feeding value.

The disadvantages of the process are the capital costs of the equipment and the fuel involved. Because of these high capital overheads, grass drying must be associated with a high throughput and a long drying season to be economically viable.

Manuring, cutting and wilting have all to be closely integrated to ensure an adequate supply of the correct type of grass to the drier.

With the escalation of fuel costs, dried grass is no longer able to compete as a bulk fodder food, but must be considered as a cheaper home-grown concentrate, comparable in price to barley and sugar beet nuts. Dried grass is now almost wholly produced as a commercial product; few, if any, farmers can contemplate grass drying on a farm scale for home consumption.

Relative Costs of Grazing and Conserved Grass

In many circumstances, the situation and size of dairy farm tend to lay down physical conditions which tend to dictate how the farm can best provide the necessary bulk food in the form of grass and conserved grass products.

Table 6.5 gives some interesting information which allows comparison of the relative costs of grass and conserved grass products.

The table gives confirmation to the idea that grazed grass is the cheapest source of energy in the summer.

Forages grown for summer grazing also have their place and

Table 6.5 Relative Costs of Grazing, Conserved Grass, etc.

	Yield DM Tonnes/ha (acre)	*Cost per tonne DM (£)*	*MJ per kg DM*	*Pence per MJ of ME in DM*
Grazed Grass	11.1 (4.5)	38.6	11.8	0.33
Kale (direct drilled)	6.9 (2.8)	42.0	11.0	0.38
Forage Turnips (direct drilled)	6.9 (2.8)	38.1	10.2	0.37
Grass Silage	11.1 (4.5)	69.1	10.9	0.63
Big Bale Silage	11.1 (4.5)	74.1	10.9	0.68
Extra Silage	2.5 (1.0)	37.0	10.9	0.34
Purchased Hay[1]	—	76.5	8.8	0.87
Brewers' Grains[2]	—	108.5	10.0	1.09
Concentrates[3]	—	155.2	12.8	1.21

[1] At £65/tonne. [2] At £24/tonne. [3] 14% CP, delivered in bulk, £133.50/tonne.

Note In interpreting the above comparative figures for use in planning feed use on the individual farm it is important to remember two points: (a) that own land, labour and capital for equipment are required for home-produced fodder but not for purchased feed, and (b) there is a limitation on the consumption of bulk fodder by ruminant livestock—although this very much depends upon its quality/digestibility.

Source: M.E. Huchinson (Henley Manor Farm) ICI Ltd, 1987. Farm Management Pocket Book, J. Nix, Wye College, 1989.

cost only slightly more than grass. Grass silage is used throughout the year and is generally twice as expensive as grass, although the cost can, of course, vary tremendously. On some farms with very sophisticated silage machinery the cost could be much higher than is shown.

The cost of purchased feeds, hay grains and concentrates are dependent on market forces at any one time. Hay and grains vary in price directly in relation to the availability of farm grown bulk food, in spring, summer, autumn and winter.

A late spring, means a late turn-out to grass, with wet conditions. Grass may be available but it is unlikely to be accessible.

A dry summer is likely to result in insufficient grass requiring the use of alternative feeds.

A wet autumn with heavy wet ground conditions may have grass available which cannot be used. The likelihood of poaching means that cattle will be housed earlier and the winter feeding period will be extended. During a dry winter, after the traumas of the spring, summer and autumn mentioned, it is highly likely that winter feed stocks could be low. Forward planning is vital. The art of buying in feed to supplement available rations is to recognise the problem at an early date and arrange contracts to organise food supplies before the demand rapidly increases as other farmers begin to realise that they also have a shortage of cattle feed.

GRASSLAND RECORDING

Many farmers grow grass but few consider it necessary to record production from grass. The recording of grassland comes into two categories:

- the effective use of grass on a day-to-day basis, using a net output method;
- the recording of grassland use over a whole season; thus allowing a comparison of husbandry systems, levels and timing of fertiliser applications etc.

These systems can be illustrated as follows:

1. Day-to-Day Grass Use

Once the grass is grown, the most important feature is to make best use of it. A fairly simple means of assessing grass use as a means of producing milk is to compare the output of milk with the input of supplementary feed over a period of time—one week is suggested. This will tend to even out daily fluctuations, so that for a 120-cow herd, largely autumn calving, out to grass in mid-July, the results might be as shown in Table 6.6.

The grass at this time was providing for Maintenance (M) + 10.42 litres milk per cow. This was a herd level of production. It must be realised that within the herd, late autumn calvers were producing 12 litres from grass whereas spring calvers would only be producing 5–7.5 litres per day from grass. The level of concentrate usage at 0.19 kg/litre and milk produced per cow are indicative of the herd production at this time of year. If this kind of assessment is carried out on a regular weekly basis, the pattern of production throughout the year can be established. The farmer can then ask himself whether it is possible to achieve a better performance the following year.

2. Grassland Assessment on an Annual (Rolling Average) Basis

Grassland recording is a technique of grassland evaluation which needs more time than the average farmer has available, certainly in the summer.

Data available in the normal type of costings on most farms can offer a simple system of measuring the efficiency of grassland use.

e.g. *Data available*—sample farm

No. of cows = 101
12 month average milk yield per cow = 6,344 litres
12 month average concentrate use per litre = 0.19 kg/l
12 month average concentrate use per cow = 1.205 tonnes/cow
Concentrates produce milk at the rate of 0.45 kg per litre of milk of average composition
Therefore 1.205 tonnes will provide 1205/0.45 = 2,677 litres
Net milk from farm forage
= total milk produced minus milk from concentrates
6,344 minus 2,677
= 3,667 litres per cow

On any farm using this simple calculation it is possible to obtain a reasonably accurate assessment concerning the production from farm resources.

There is a very wide range of values possible from 0–4,000 + litres.

A town dairy situation where everything is purchased, even the maintenance foods, would produce the lower figure. Such a farm is, however, unlikely to be found currently, except perhaps where there has been urban development or in a village situation.

The opposite end of the scale is likely to occur where land is unlimited. With stocking rates at 1.8 cows per hectare, the low input/low output system could be practised and milk from forage could well exceed the 4,000 litre mark. However just to illustrate that high output can be associated with high forage use, a recent UK dairy herd prize winner had a yield of 8,000 litres on a 100 cow herd. Concentrate usage was at 0.23 kg/l. and milk from forage was 4,000. The farmer incidentally won the National Silage competition in the same year. The real answer then is making the best possible silage and using it to the best possible advantage.

Table 6.6 Production Pattern

Total milk from herd during the week:	9,711 litres
Total concentrates fed during the week:	1,875 kg
(high energy concentrates at 13.5 MJ/kg DM need to be fed at 0.45 kg per litre)	
Hence milk from concentrate:	$\frac{1,875}{0.45}$ = 4,166 litres
Milk from grassland	
Total milk:	9,711 litres
less milk from concentrates:	4,166 litres
equals	5,545 litres from grass
Milk from grass per cow per day:	$\frac{5,545}{\text{No. of cows in milk (76)} \times 7}$ = 10.4 litres
Milk produced per cow per day:	$\frac{9,711}{\text{No. of cows in milk (76)} \times 7}$ = 18.2 litres
Actual concentrate usage:	$\frac{1,875}{9,711}$ = 0.19 kg/litre

Utilised Metabolisable Energy—UME

This is an objective method of measuring output of grassland.

It is possible to describe it in simple terms as the contribution made by grass both grazed and conserved to the total energy requirements of the ruminant.

This can be set against the expectations for the production of grass and apparent efficiency of grassland utilisation calculated for the farm. To calculate the UME it is necessary to know:

(i) The number of livestock, weight and level of production
(ii) The quantity and quality of all purchased feeds
(iii) The grassland area devoted to livestock enterprise

This would involve the detailed recording of the use of all fields by the appropriate classes of ruminants on the farm. As previously mentioned these records are to be found on very few dairy farms, because of the amount of time it would take. UME is measured in units of energy known as gigajoules (1,000 megajoules), and expressed either in GJ per cow or per hectare.

CHAPTER 7

THE SCIENCE BEHIND FEEDING COWS

Scientific feeding of dairy cows had been practised for many years.

The basis for scientific feeding is the understanding that animals have two functions to perform:

(i) *Maintenance:* This involves keeping all the bodily processes going and keeping the animal in good health without any weight loss.

(ii) *Production:* Any nutrients in the ration which are surplus to those needed for maintenance will be available for either: growth in young animals or increasing live weight gain in mature animals; the production of milk in the lactating cow, sow, ewe or mare; or the performance of work, as in the case of horses and other work animals, e.g. draught oxen in undeveloped countries.

FEEDING THE DAIRY COW AND THE ME SYSTEM

The method of expressing food requirements has been the subject of much discussion and thought in the United Kingdom for many years. In 1975 a joint working party of researchers, advisers and teachers produced a report, *Energy Allowances and Feeding Systems for Ruminants*—Ministry of Agriculture Technical Bulletin No. 33, which was adopted in the United Kingdom, replacing the previous Starch Equivalent System. The new system was concerned with the use of Metabolisable Energy (ME) as a basis for formulating rations on the farm.

Feeding dairy cows can be described in the simplest way by matching up two basic factors:

- What does the cow need?
- How can farm foods meet this requirement?

The chief limiting factor in the food supply to the dairy cow is energy. This value may be expressed in various ways, but the system adopted in the UK is that outlined as in 'Bulletin 33'.

To standardise the measurement of energy, it must be measured on a dry matter basis. The water content of foods is variable and does not contribute to the energy level of the food. For example, if a silage sample is said to have 25 per cent DM—this would mean that 100 kg of silage would be made up of 25 kg of DM and 75 kg of water. Hay would be expected to have a DM of 85 per cent.

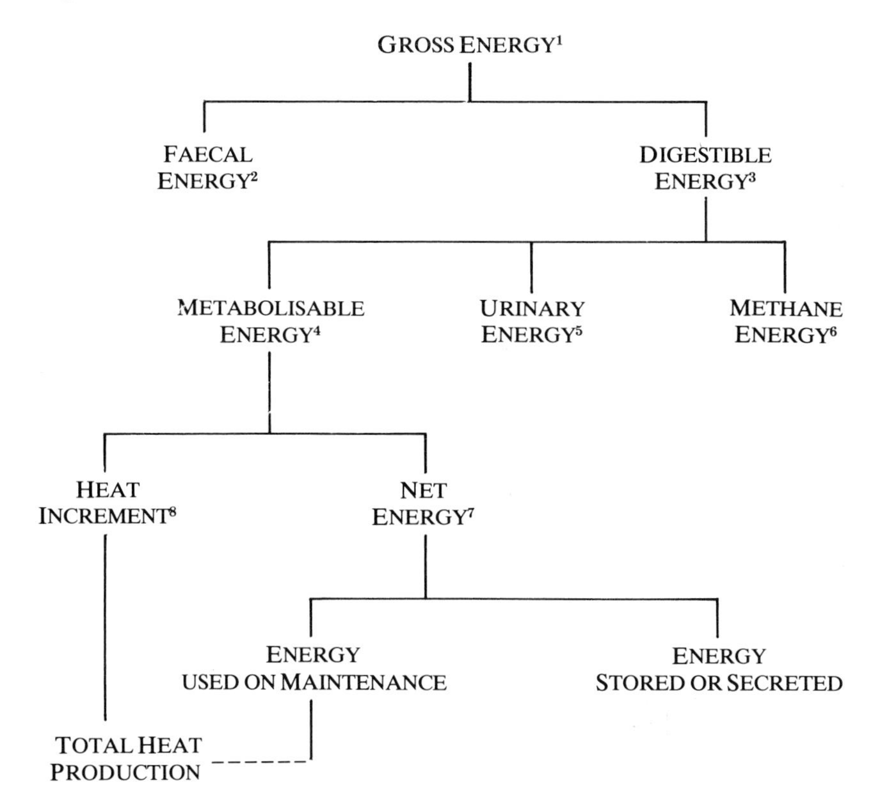

Figure 7.1 The background to the ME story
Energy Allowance and Feeding Systems for Ruminants Technical Bulletin 33, MAFF.

The utilisation of energy in food is illustrated in Figure 7.1. The various energy categories can be defined:

1. Gross energy—total energy in a feed.
2. Faecal energy—energy lost in the faeces (dung).

3. Digestible energy—the energy left which can be further used by an animal.
4. Metabolisable energy—the energy extracted by the animal from the feed to be used in maintenance and production—due allowance must be made for the loss of energy in the urine (5), and the fermentation process (6), which results in the production of methane gas. The cow disposes of this when she ruminates or belches.
7. Net energy—the collective name for the energy used for maintenance and production.
8. Heat increment—is the amount of energy used up by converting the ME into maintenance and production; also for the important digestion, transport and basic body processes like body temperature, blood circulation etc.

The basic unit of energy in the ME system is called the megajoule (MJ).

The requirements of metabolisable energy for maintenance and production are supplied in Table 7.1 (page 127).

One of the factors that was not previously considered in rationing dairy cows was the loss and gain of liveweight through lactation. This has been taken care of by the ME system. For 1 kg of liveweight loss there is a contribution of 28 MJ of dietary ME, and for a gain of 1 kg liveweight, the ration must supply an extra allowance of 34 MJ.

There is a range of energy and protein allowances between and even within breeds. Generally speaking, better feeding allows dairy cows to attain their genetic potential in terms of quality, and this aspect is in increasing importance since payment depends increasingly on quality.

FEEDING COWS—THE PROTEIN STORY

For many years, farm foods have been analysed and the protein part of the food classified as Crude protein (CP) and True protein (TP). The main difference between the two being the non-protein nitrogen (NPN) in the food which would reflect some mid-way point between an immature product like spring grass and a mature fibrous product such as hay.

The protein part of a food was further evaluated as being digestible to the animal—Digestible Crude Protein (DCP). The DCP approach was to regard the ruminant as an animal with only one stomach and took no account of the ruminants ability to utilise non-protein nitrogen. A new protein system was recommended in 1980 by the Agricultural Research Council based on protein research with ruminants.

The very important first step however, is to realise that the energy level of the ration must be adequate to meet the desired level of production. Once this energy level is reached then the next step is to provide other nutrients including protein, so that the production potential can be realised.

The protein required to help achieve the required production level is now known to be available in two forms:

1. Rumen Degradable Protein (RDP)
   This is the protein component of the food which is released when the food is partially digested in the cows rumen (the first stomach). Some of this protein is used by the microbes themselves for their own protein needs. Any excess degraded nitrogen will be absorbed from the rumen, converted to urea in the liver and then removed by the kidneys as urine.

2. Undegradable Protein (UDP)
   This is protein which is not digested in the rumen. The synthesised microbial protein and other unfermented feed protein

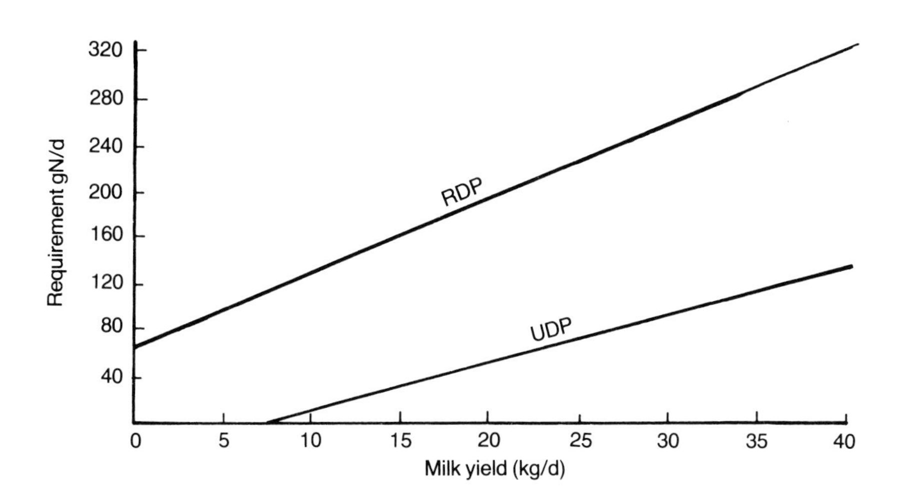

Figure 7.2 Increasing requirements of the dairy cow for RDP and UDP with milk yield
Improved Feeding of Cattle, Wilson and Brigstoke

(UDP) pass into the abomasum (the fourth stomach) where they are broken down into amino acids which are necessary for milk production and liveweight gain.

Work by the Agricultural Research Council includes the classification of foods into the RDP/UDP ratios and from this it is possible to match the various foods of differing degradabilities according to production requirements.

The Ratio between RDP/UDP

Research work suggests that where animals are giving low yields of milk, at the rate of 7.5 litres per day, the protein can be provided by RDP alone. Yields above this level need contributions from both RDP and UDP. Where only RDP was provided, the only source of nitrogen was a synthetic ration which had urea as the nitrogen source.

Figure 7.3 illustrates how the degradability of the ration must decrease with increased milk yield, in a similar way that energy concentration increases with higher yields. With higher milk yield it is important for protein to avoid the degradation process in the rumen and to pass into the final gut for absorption still in the amino acid state.

The ratio of RDP/UDP is perhaps sufficiently complicated for it to be worth dairy farmers seeking technical help in getting the correct balance worked out for their particular herd requirements.

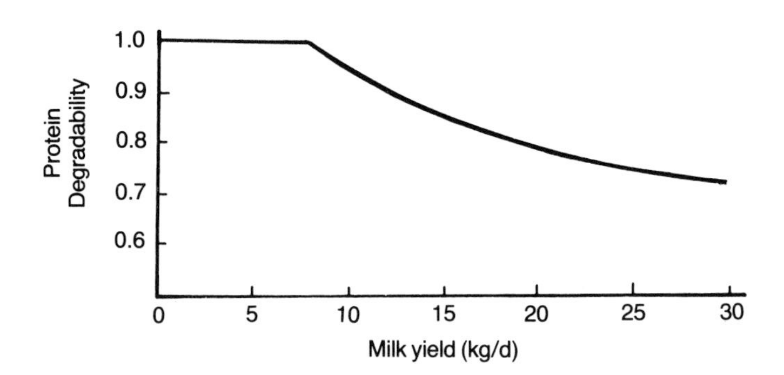

Figure 7.3 Relationship between protein degradability and increasing milk yield
Improved Feeding of Cattle, Wilson and Brigstoke

DRY MATTER INTAKE

Having calculated the ME and DCP requirements of a cow, the last basic consideration to take into account is the amount of feed that an animal will eat. This could be difficult when one takes account of the varying dry matter of different types of foods. However, the calculations are simplified by having all food intakes related to the dry matter intake of the cow. Table 7.1 shows how the cow's appetite in terms of dry matter intake per day is related to bodyweight and also to the amount of milk being given.

In the case of high-yielding cows, it is a fact that in practice, such cows often have the ability to eat far more than the theoretical amounts of dry matter.

Table 7.2 calculates an example ration which supplies the necessary nutrients for energy, protein and dry matter intake.

The ration in Table 7.2 has been calculated using the generally accepted nutrition tables produced by the Ministry of Agriculture, Fisheries and Food (Table 7.1.)

From the ration, one can see that the calculations are quite precise (to two places of decimals). In the practical feeding situation however, the dairy parlour feeders on many farms will only feed to 0.5 or even 1 kg. The main purpose of the calculation, therefore, is to appreciate the balance of foods, bulk and concentrates, so that every effort can be made to obtain milk production with as big a margin over feed as possible. The final answer in the calculated ration is:

50 kg of silage
2 kg of sugar beet pulp
5.5 kg of medium energy 14 per cent protein concentrates

Considering the slight excess of DCP, there could be fine tuning exercised to the extent of exchanging 1 kg of sugar beet pulp nut for 1 kg of protein concentrates. Feeding extra silage would also allow regular reduction in concentrate levels.

Once the ration is finalised and being fed, the question arises as to whether the ration could be cheapened.

Steps have been taken already to cheapen the concentrate part of the ration by including sugar beet. This would probably cost between £100–110 per tonne compared to £135–140 per tonne for a proprietary 14 per cent medium energy concentrate.

Table 7.1 Nutrient Allowances for Dairy Cows
(a) For Maintenance

	Live weight	*Typical appetite*	*Daily allowances for maintenance*					
		DM	*ME*	*DCP*	*Ca*	*P*	*Mg*	*Na*
Breed	*kg*	*kg*	*MJ*	*g*	*g*	*g*	*g*	*g*
Jersey	400	16	45	250	14	19	7	3
Guernsey	500	16	54	300	18	25	9	4
Ayrshire	550	17	58	325	20	29	10	4
Shorthorn	550	17	58	325	20	29	10	4
Friesian	600	18	63	345	21	33	11	5
Holstein Friesian	625	19	65	355	22	34	11	5
South Devon	650	20	67	365	23	35	11	5

(b) For Maintenance and Pregnancy

	Live weight	*Daily allowances for maintenance and pregnancy for 150 days before calving*					
		ME	*DCP*	*Ca*	*P*	*Mg*	*Na*
Breed	*kg*	*MJ*	*g*	*g*	*g*	*g*	*g*
Jersey	400	68	370	30	27	10	5
Guernsey	500	77	430	33	33	12	6
Ayrshire	550	81	465	35	37	13	7
Shorthorn	550	81	465	35	37	13	7
Friesian	600	86	495	37	41	14	8
Holstein Friesian	625	88	515	38	42	14	8
South Devon	650	90	530	39	43	15	8

(c) For Milk Production

Metabolisable energy					*Other nutrients (g/kg milk)*	
Milk fat content %	3.5	4.0	4.5	5.0	DCP	60–70 (Early lactation) 55–65 (Mid-lactation)
ME allowance (MJ/kg milk)	5.0	5.4	5.7	6.0	Ca	2.8
					P	1.7
					Mg	0.7
					Na	0.6

Source: ADAS MAFF Publication. P2087, 1988.

Table 7.2 Daily Ration For a Holstein/Friesian Cow of 625 kg Liveweight With Reference to DM, ME and DCP Requirements

Production details
Calved in October, recorded mid March at 25 litres, 4% BF.

Foods available	*% DM content*	*ME MJ/kg DM*	*DCP g/kg DM*
Silage	25	10	90
Sugar beet pulp nuts	90	12.5	80
Concentrates medium energy	86	12.5	140
Requirements of the cow			
At 625 kg liveweight	19	–	–
For maintenance	–	65	355
For production	–	135	1375
Total requirements	19	200	1730
A possible ration would be:			
50 kg of silage (as fed)	12.5	125	1125
DM 50×25%=12.5 kg DM ME 12.5×10=125 MJ DCP 12.5×90=1125 g DCP			
2 kg of sugar beet pulp (as fed)	1.80	22.5	144
DM 2×90%=1.8 kg DM ME 1.8×12.5=22.5 MJ DCP 1.8×80=144 g DCP			
Sub total	14.30	147.5	1269
Shortfall in DM (19–14.3=4.7) This is made up with concentrate= ME=4.7×12.5=58.75 MJ DCP=4.7×140=658g DCP NB: 4.7 kg DM concentrates=5.5 kg concentrates as fed.			
Shortfall made up by	4.7	58.75	658
Total nutrients provided	19.0	206.25	1927
Total requirements	19.0	200.00	1730
Possible surplus/deficit	—	+6.25	+197

The average cost of the concentrate part of the ration would be

2 kg of SBP @ £105/tonne (10.5p/kg) =	21p
5.5 kg of concentrates @ £140/tonne (14p/kg) =	77p
Total 7.5 kg	98p

Average cost per kg $= \frac{98}{7.5} =$ 13p per kg or £130 per tonne.

Viewing the concentrate cost with an enquiring mind may allow further economies to be made.

The biggest area for variations in the ration is the question of dry matter appetite. Ministry leaflets, comment that appetite may vary by as much as 20 per cent in individual animals according to stage of lactation or pregnancy, in addition to the forage variables of the diet's energy concentration (M/D*) and dry matter content.

The calculated ration is the starting point, from which deviations can be introduced. The need for modification becomes very obvious when the ration fails to produce results in line with the predicted levels of milk production. In such an event, time must be taken to identify the fault and apply corrective action.

IMPORTANCE OF MINERALS

The mineral content of the diet is important. With farm-mixed foods a mineral mixture needs to be incorporated. There are many commercial mixtures available to meet any situation. Most oil cakes are rich in phosphorous relative to calcium, whereas most grassland products (hay, silage, dried grass) and beet pulp are rich in calcium and low in phosphorus. Compound cakes are fortified by the necessary minerals.

* M/D—The metabolisable energy concentration of a ration per unit of dry matter (measured in MJ/kg DM or simply MJ/kg).

e.g.	12 kg DM silage @ 10.5 MJ/kg DM =	126
	4 kg DM concentrates @ 12.5 MJ/kg DM =	50
	16 kg	176

$$M/D = \frac{176}{16} = 11 \text{ MJ/kg DM.}$$

Gross mineral deficiencies are unlikely to be found on farms where good husbandry is practised, even when production is obtained largely from home-grown non-concentrate foods. In some areas trace-element deficiencies (e.g. copper) are known to exist.

The indiscriminate addition of minerals to a cow's diet is to be deprecated. When in doubt, veterinary advice should always be obtained.

The chief mineral deficiencies occurring today in order of frequency are as shown in Table 7.3.

Table 7.3 Mineral Deficiencies

Condition	*Deficiency*	*Prevention*
Lactation tetany	magnesium	Add 57 g per day calcined magnesite to concentrate ration
Pining Failure to thrive Dull coat	cobalt or copper	Feed suitable reinforced mineral mixture under veterinary advice
Goitre Swollen glands Infertility	iodine	

VITAMINS

On the farm, the most likely symptoms of mineral deficiencies are indicated by a depraved appetite and skeletal weakness in milking cows. If the assimilation of lime and phosphorus into the blood stream and hence to the skeleton is to be efficient, the provision of adequate minerals in the diet of milking cows or young calves must be associated with adequate supplies of Vitamin D. The cow and the calf can both manufacture their own Vitamin D under the effect of direct sunlight, but in winter when they are housed, Vitamin D deficiency can be avoided by feeding them good sun-cured hay, or cod-liver oil or synthetic preparations of Vitamins A and D in powder form. Usual rate of addition is 2.2 kg synthetic vitamins mix per tonne of concentrates.

The direct addition of vitamins to the cows' diet—except in respect of Vitamins A and D (see above)—is not usual or necessary provided the diet is composed of a basis of good-quality home-grown foods, particularly hay, silage and dried grass. Green foods, providing the yellow colour in the milk, have a high β-carotene content—the precursor of Vitamin A which has an important function of its own in relation to fertility.

SCIENCE RELATED TO PRACTICE

The feeding standards illustrate the variations in the food requirements of dairy cows both in respect of the breed and the quality of the milk produced. To use them in practice it must be remembered that dry matter intake by the cow is governed by appetite and the palatability of the foods fed. Mention has been made of the varying moisture content of feeds. In practice, there are many simple features which have to be borne in mind when making up rations. Sugar beet pulp, for example, swells when moistened and gives the cow a feeling of repletion before the required dry matter intake has been consumed. Furthermore, food intake per day is dependent on the speed at which it passes through the cow and is digested. On a laxative diet, possibly spring grass, the food is low in fibre and passes through the animal so quickly that the animal cannot extract enough energy to satisfy its needs. On a high-fibre diet, the throughput of food is much slower; the fibre is associated with nutrients being indigestible and once again the animal obtains less than her requirements of energy and protein.

A SILAGE GUIDE

Silage is a widely used food on dairy farms. The question of appetite and value of the ration becomes particularly important when deciding on the level of silage to be included on the diet. Silage is usually made in three cuts and there can be very different dry matters in each cut. As the cow progresses from one cut of silage to the next this change in dry matter content must be taken into account.

In summary it can be said that the appetite of a cow is affected by:

- The production level and general rate of metabolism in the cow. High levels of production raise the appetite; ill health lowers it.

- Digestibility. Foods of low digestibility depress the appetite.
- The size and relative digestive capacity (i.e. size of rumen) of the individual cow.

Table 7.4 A Guide to the Dry-Matter Content of Silage

Physical test	*Percentage dry matter*
Water easily squeezed out by hand	under 18
Water just expressed by hand	19–24
Water not expressed by hand	over 25

VALUE OF HOME-GROWN FOODS

With the knowledge of cows' food requirements, their digestive capacity and the feeding value of foods available, it is possible to feed along scientific and economic lines. The herdsman's real problem is determining the value of home-grown foods. Purchased feeding stuffs, bought under guaranteed analyses as required under the Fertiliser and Feedingstuffs Act, will vary very little in compositional quality. Home-grown feedstuffs on the other hand vary considerably in value and usually make up a major part of the cow's diet. It is vitally important, therefore, to be able to assess the feeding value of home-grown foods with reasonable accuracy.

The answer concerning the value of a diet must come inevitably from the cow herself. For years there has been too little correlation between the make-up of a ration and whether it achieved its desired objective. When a ration does not produce the desired result it is highly likely that the fault lies in the make-up of the ration rather than with the cow herself. This emphasises the need to set up regular monitoring systems giving a detailed report of the relationships between milk output and feed input.

Stage of Growth Affects Digestibility

The feeding value of grass and its conserved products varies with the stage of growth at which it is eaten or conserved. Increasing age and a slowing down of the growth rate, are accompanied by a falling percentage of protein and a rising percentage of fibre. An increase in the amount of fibre consumed lowers the efficiency of digestion and forces a cow to work much harder in order to extract the available food nutrients.

This is of little significance with low-yielding animals, but when high levels of production are the aim, the fibre intake becomes important.

Many home-produced foods are high in fibre. For example, most samples of hay will be within the range 44 to 48 per cent fibre with dried grass and silage varying from 25 to 40 per cent; whereas, in contrast, a typical compound cake or meal has from 8 to 10 per cent. It is essential, therefore, to know the stage of growth at which hay, silage or dried grass are cut. The earlier the cutting is taken, the lower is the fibre percentage.

More Fibre—Lower Value

Fibre is rather a loose term and difficult to define chemically. It comprises cellulose and hemi-cellulose, both readily digested unless tightly bonded to lignin (woody tissues) which in itself is non-digestible.

The make-up of the sward cut is important since certain grasses with a high stem-to-leaf ratio give rise to fibrous material. Loss of leaf in haymaking also results in a serious diminution of feeding value.

This lowering of feeding value with maturity is also seen in certain root crops such a marrow-stem kale, which tends to lose its leaves and become increasingly fibrous after mid-winter. On the other hand, mangolds and swedes contain 12–18 per cent of fibre in the dry matter, and therefore form excellent companion foods to feed with the fibrous cereal straws which contain 40–50 per cent fibre.

Basically, therefore, high-quality hay, silage or dried grass can only be made by conserving the right type of material at the right stage of growth.

A further factor of conservation is that of palatability. Well-made hay or silage is highly palatable. On the other hand if hay is badly weathered or silage incorrectly fermented, then moulds develop resulting in a corresponding loss of feeding value and palatability. Less of the food is eaten, and more wasted.

Stockmen should also note that dirty feeding troughs, or soiling of the food, for instance with soil or faeces—as often happens with sugar beet tops—greatly reduces palatability.

Urea-based supplements are available in liquid or solid form. The photo on the left shows a specially designed feeder for the liquid supplement; the bottom picture shows an animal taking block supplement from a steel tub.

Getting an early bite from a ley top-dressed with nitrogen in February. To counteract the laxative nature of young grass and to mitigate the risk of bloat, the changeover from winter feed to summer grazing should be made gradually.

Grazing a ley during late summer. In practice, the production value of late season grass is at least 10 litres less per cow per day than similar grazing in spring.

A Food Classification

In order to concentrate the diet of high-yielding cows and render that diet readily digestible, we need to use feedingstuffs high in digestible nutrients and low in fibre. Such foods are commonly known as concentrates and include cereal grains, oil cakes and so on. The available foods can be classified as in Table 7.5.

Table 7.5 Classification of Feedingstuffs in Ascending Order of 'Concentration'

Group A	
Roughages (low concentration)	
Fibre percentage 35 per cent or over	*Remarks*
Cereal straws (oats and barley only)	Derived from mature
Hay (average or below)	plant tissues, suitable
Low-quality silage	only for 'maintenance'
Low-quality dried grass or lucerne	purposes
Group B	
Succulents and high-quality roughage (medium concentration)	
Fibre percentage 15–35 per cent	*Remarks*
Grass and arable silage	Suitable for maintenance
Root crops	and for low-level
Kale and sugar beet tops	production
Group C	
Concentrates (high concentration)	
Fibre percentage below 15 per cent	*Remarks*
Cereal grains and dredge corn	Essential foods for high
Milling offals	production
Distillers' and brewers' grains	
Oil cakes and meals	
Meat and fishmeals	

It should be noted that high-quality dried grass is classified as a concentrate only when its fibre content is below 20 per cent, irrespective of its crude protein content.

With silage having a good level of protein, it has been usual to feed cereals or sugar beet pulp for the first 5–10 litres of milk on top of the silage on a feed fence.

COMPOUND CONCENTRATES

The majority of milk production is obtained by feeding one or other of a wide range of commercially compounded foods. They are available in several categories.

High-Energy Compounds

This type of compound has an ME of approximately 13.5 MJ/kg DM. It is often associated with a high level (around 18 per cent) of protein. In the majority of herds, the main fault with the rations fed is the shortage of energy.

Using the ME system, it can be shown that high energy concentrates with 13.5 MJ/kg DM should be fed at 0.45 kg/litre to provide adequate energy for milk production (see calculation below). The higher levels of protein are also associated with and essential for the higher yielding cows.

High-energy compounds usually contain a higher percentage of oil with 6–7 per cent being the maximum to maintain good physical quality, i.e. to produce a hard nut which will not disintegrate into dust. Too much oil interferes with digestion in the rumen.

Concentrate Feeding for Milk (High Energy)
1 kg concentrate DM contains 13.5 MJ. 1 kg concentrate (as fed) at 86 per cent DM contains $13.5 \times 0.86 = 11.61$ MJ. A litre of milk at 3.9 per cent BF needs 5.2 MJ.

Hence 1 kg of concentrates (as fed) is sufficient for $\frac{11.61}{5.2} = 2.23$ litres.

Rate of feeding required: kg per litre $= \frac{1}{2.23} = 0.45$

Medium-Energy Compounds

This type of food has an ME of 12.5 MJ/kg DM (approx. 14–16 per cent protein) and will normally have to be fed at a rate of 0.48 kg per litre. This type of compound will usually be used in herds up to the 5,400 litre level. Higher yields require higher-energy concentration in the ration, i.e. a higher M/D.

Modern concentrates are made up on a least-cost formulation system so the actual constituents do vary from time to time.

In the fifties and sixties there was a great deal of interest in the home-mixing of bought-in concentrate foods for the production ration. A combination of several factors has had the effect of reducing much of the home mixing:

1. With herds increasing in size, the labour on many farms has been matched to the efficient running of these larger units, and labour has not been available to carry out the mixing.
2. In recent years the amount of capital investment necessary in new milling and mixing plant has been such as to reduce the interest in the project. Mobile mill and mix units are now available in most areas.

On grassland farms in the west, where there has been a move to silage, this has been widely supplemented by bought-in balanced concentrates for the greater part of production.

The choice between buying in concentrates or home mixing is very much related to the type, size of farm and all the resources available, particularly labour and capital.

Use of Straights

Straights is the common term applied to single concentrates which may be either purchased or home grown and are used to replace proprietary dairy compounds which inevitably are more expensive. Straights include barley, wheat, dried sugar beet pulp, maize gluten meal, dried distillers' grains, soya bean meal and palm kernel cake.

Their use has become increasingly popular as the farmers' attention has been focussed on the importance of margin per litre. A number of factors have combined to encourage farmers to use more straights.

(i) The price of straights is considerably lower than compound feeds. Mention has been made of sugar beet at £100–110 per tonne compared to the wide range of compounds ranged in price from £135–170 per tonne. If a feed is cheaper, and can help to produce the required milk yields, then there will be interest in its use.

(ii) With the continuing emphasis on making good quality silage, so the extra quality can not only fulfill the maintenance requirement but also contribute considerably to milk production itself.

Good quality silage, similar to grass, will always have a surplus of protein beyond the ration requirements. This can be translated into increased margin per litre if the extra protein is matched up with an energy rich food; sugar beet nuts or rolled barley are obvious partners.

Where the silage quality is lower, with a reduced protein level, it is still possible to provide for at least 10 litres of milk production from a combination of 2 kg of sugar beet/barley and 2 kg of a protein-rich 'straight' e.g. maize gluten meal or dried-distillers' grains.

Single feeds or mixtures such as those described are very suitable when a herd is grouped, with the high yield section having a third feed per day on the feed fence. Supplementary 'straights' can be put in a feed fence on top of silage or mixed with silage in a forage box.

The main attraction is the cheaper cost. When maintenance feeds are in short supply and alternatives are required, then the comparison should be made by equating the feeds on a DM basis and then comparing unit cost. For example:

Silage at £22 per tonne at 22 per cent DM: ME = 10.2 MJ.
A 'straight' concentrate at £100 per tonne at 86 per cent DM: ME = 12.5 MJ.

(i) It would need 4 tonnes of silage to equate to 1 tonne of straights on a DM basis.

(ii) Silage—4 tonnes at $\frac{£88}{10.2}$ = £8.62 per unit MJ

Straight–1 tonne at $\frac{£100}{12.5}$ = £8.00 per unit MJ.

NB. DMs available may vary in practice from 18–30%. With practice a rough assessment can be made by squeezing some of the silage in your hand.

Availability and price per tonne will constantly vary according to pressure of demand. The calculation referring to unit cost on an equal DM basis will always give an accurate picture of the comparative values.

Research into Ruminant Digestion

Recent work has focused attention on the importance of the 'physical texture' of food consumed by the cow in its effect on the production of fatty acids by bacterial fermentation in the rumen or first stomach.

Roughage foods tend to promote acetic acid production, whereas concentrates tend to promote propionic or butyric acid production. Acetic acid encourages butterfat, whereas propionic or butyric acid encourage higher solids-not-fat production, so that the balance of roughage to concentrates in the diet of the cow can influence milk composition. Roughage foods which are finely ground (i.e. the fibre is no longer in the long state) act similarly to concentrates. In order to encourage butter fat production it is therefore desirable to include some long fibre in the diet.

HOUSING AFFECTS FEEDING

It must not be assumed, however, that the previous tables cover all the variations in maintenance and production requirements. In fact, they refer to a fairly standard set of conditions—namely those which occur when the cow is housed.

In practice, variations from these standards generally tend towards increasing maintenance requirements. Thus, when grazing, the cow uses up energy searching for her food, especially on thin, low-density swards or where long distances have to be travelled to water or pasture. Where cows are not housed, exposure to wet weather increases the heat loss from the body and raises the amount of maintenance food required.

No advantage is gained by keeping cows too warm: 10–13°C is about right. But cows should be able to lie down comfortably to rest and should not be forced to stand because of fear of aggressive companions in yards or adjoining stalls, or because of adverse conditions outdoors.

In regard to production requirements, it is not suggested that the above standards apply under all conditions. With high-yielding cows, or when feed prices are low relative to milk prices, it may well pay to feed slightly above the level of these feeding standards, but the increase in milk output for each increase in food consumed is subject to the law of diminishing returns which sets a ceiling on what is a profitable feeding level.

THE COW KNOWS BEST

One last fact must be noted. Feeding standards must inevitably relate to the average cow, and therefore may not fully meet the actual feed requirements or appetite of the individual cow.

In general, the higher the genetic milk potential of the cow, the greater is her digestive efficiency. If properly fed, she will produce milk more economically than the lower yielder, both in terms of energy and protein consumed. In most cases, she will also show a greater dry-matter consumption than the average cow, though 'shy feeders' with a high productive capacity are not uncommon.

Rigid adherence to feeding standards represent a pedantic approach to the problem. The wise and observant stockman is prepared to learn from his cows as to whether his theoretical estimation of their requirements is in fact producing the expected results. This can only be assessed if the quantities of food fed are known and there is a factual or objective appraisal of the cow's performance and condition.

COW CONDITION IS IMPORTANT

Is the milk yield satisfactory? What change is occurring in the condition of the cows? These are questions a herdsman will continually be asking. Loss of flesh, for example, can be an indication of underfeeding, not perhaps in total food, but in actual nutrients consumed. For a short period, loss of flesh may be sound husbandry, but this condition should not be allowed to proceed so far as to impair the health or stamina of the cow, her breeding regularity or the quality of her milk.

CHAPTER 8

PLANNING THE BULK FEED SUPPLY FOR THE DAIRY HERD

WINTER FEED PLAN FOR THE DAIRY HERD

With the knowledge of the basic requirements for dry matter, energy and protein to both maintain the cow and produce milk, it is then possible to attempt a calculation concerning the total requirements of a dairy herd for a winter period.

Table 8.1 gives the outline of a scheme for working out the amounts for a group of Holstein/Friesian cattle, in a herd producing an average of 5,400 litres at 4 per cent butterfat. When working out a calculation of this nature it is necessary to make a number of, hopefully, reasonable assumptions. It is also helpful to cut a few corners, such as assuming that all months have 30 days.

An average animal from an October-calving group, would be likely to give the yield shown in the table. Silage can vary in quality over a very wide range. In this case, it is assumed to have an average of 25 per cent DM, and is expected to have sufficient nutrients to produce maintenance plus 5 litres in early lactation, increasing to virtually all production by grass turn-out time in April. No concentrates would be fed to this group after turn-out.

The value for dry matter appetite is appropriate for the breed mentioned. i.e. 625 kg liveweight. It should be noted however, that DM appetite can vary by as much as ±20 per cent depending on the physical nature of the bulk feed.

During the winter, yield decreases, concentrate usage decreases and the amount of silage eaten increases.

This will be different for each monthly calving group of cows, so the detailed answer will vary in respect of the amounts of silage and concentrates eaten. Some monthly groups will be dry for 2 months in the winter.

Information is limited concerning the appetite of the dry cow; the DM appetite is said to be lower but this is difficult to

Table 8.1 Assessment of Total Silage and Concentrate Use For 210 day Winter (with October-calving cows)

Month	*Oct.*	*Nov.*	*Dec.*	*Jan.*	*Feb.*	*Mar.*	*Apr.*	
Yield litres	20	26	24	22	20	18	16	
Assessed silage value	M+5	M+5	M+5	M+7.5	M+10	M+12.5	M+15	
Concs. fed (kg DM)	5.7	8.1	7.3	5.5	3.8	2.0	1.0	
DM appetite (kg)	17	19	19	19	19	19	19	
Silage required (kg DM)	11.3	10.9	11.7	13.5	15.2	17.0	50% silage[a] 50% grass	
As fed (kg) Silage 25% DM	45	43	46	54	60	68	36	Total
Silage per month [b] (tonnes)	0.675	1.300	1.300	1.600	1.800	2.000	1.000[a]	9.675
Concentrates per [b] cow per month (kg)	100	283	256	195	114	60	minimal 15	1023

[a] + grass buffer feed.
[b] First month of lactation, cows calving throughout month—average of 15 days.

For a 210-day winter, silage required: 10 tonnes approximately.
concentrates required: 1 tonne approximately.

verify. These questions, however, are not too important since the real object of the exercise is to calculate how much feed is required to keep the dairy cow through a 210 day winter and ultimately, of course, to make sure that the feed supply is totally adequate.

It is all summed up in the expression 'There are no prizes for running out of silage in February'.

With concentrates, the problem is minimal—any difficulties are simply resolved by picking up a phone and ordering the appropriate amount for the ensuing month.

The Table suggests that the silage requirement would be in the order of 9½ tonnes per cow. All dry cows could be fed to appetite with silage, and consumption for the winter would not be too different from the example. It is therefore good planning to assume that the amount per October-calved cow should be applied to the total number of cows in the herd for the winter.

An estimate of available silage should be made very soon after the second cut is completed, allowing any shortfall between winter requirements and silage availability to be identified.

Shortages can be covered in numerous ways.

(i) A third or even fourth cut of silage can be made as required. Any surplus grass growing must be conserved and not wasted.
(ii) Steps can be taken to drill Italian ryegrass or rye for an early bite. The crop must be drilled in early September to ensure establishment and eventual very early spring growth.
(iii) The third possibility would be to buy in appropriate bulk feed, e.g. brewers' grains, pressed sugar beet pulp, or stock feed potatoes. There are often good 'buys' available in the summer before average farmers realise the possibility of winter feed shortages.

Fresh brewers' grain and pressed pulp become very scarce into the early New Year, so it is better to feed bought in supplies during Oct.-Dec., leaving the farm produced bulk for the second half of the winter.

Summer Feed Plan for the Dairy Herd

The essence of any plan for feeding the dairy herd in the summer is the policy of 'planning ahead'. Changes should be anticipated and evasive action taken before things go wrong.

An average stocking rate is about 2.0–2.2 cows per hectare. (1.1–1.2 acres per cow). The British climate is such that the production from a hectare can vary to as much as 30 per cent more or less than average in any one summer.

The following points are important when producing a summer feed plan.

(i) Early fertiliser applications are important to get the grass away to a good start. Stocking rate from turn-out (1 April–1 May) to late June can be varied from 0.13–0.2 hectare/cow, (0.33–0.5 acres/cow). Grazed grass is the cheapest feed available so utmost use must be made of it.

(ii) Grass must not be allowed to become mature. When areas of grass are getting too mature they must be cut and conserved. For small areas, the big bale system is useful.

(iii) The quality of grass varies from very lush, to very mature and fibrous. Supplementary feeding of silage is a valuable technique. When silage is put out, the cows are given the opportunity to tell the farmer whether there is sufficient grass to satisfy their DM appetite. Third cut (August made) mature grass silage is a marvellous supplement for lush May grass.

The practise of supplementary feeding could be extended to include the other crops which are fed, e.g. early rye/perennial rye grass, stubble turnips, kale, zero-grazed forage maize.

The overall intention is to supply any deficiency of either quality or quantity in the basic maintenance ration from early spring through summer to autumn. The forage crops also fit into the pattern of sowing maiden seeds as part of the overall policy of ley farming. Buffer feeding of silage is known to improve both yield and milk quality. The trend has been towards feeding big bale silage. The system is very flexible, which is an essential feature considering how climatic conditions can affect grass supplies within a very short space of time. The amount of silage to be set aside for summer use is however difficult to estimate.

It was mentioned before that all grass which grows should be utilised, i.e. 3rd and 4th cuts of silage made whenever possible.

A useful phrase to remember is that 'you can never have too much silage'—if there is a massive surplus then it can always be sold.

Assuming a grazing period for most dairy cows of 150 days at 6 kg per day. This would approximate to 1 tonne per cow for buffer feeding through the summer. Recent dry summers have pushed the requirement up to 2 tonnes/cow.

(iv) After the silage cuts are made, the silage aftermaths become available for grazing, reducing the pressure on the rest of the grazing area. Fertiliser applications should be maintained as necessary, except, perhaps during a lengthy summer drought.

Nitrogen application will produce an excess of September/October grass which the newly calved autumn cow cannot cope with. Such grass should be conserved, since it will make an acceptable silage of relatively low DM to supplement very dry grass the following summer. The on-going feature of feeding cows at grass must be the regular recording of production to identify any abnormal yield changes which occur. These are the warning lights which indicate that whatever grass is being offered is not good enough, either in quality or quantity, to do the job required.

(v) Buffer feeding. This is a very good name for a feeding practice that has been used by dairy farmers for many years. It could be described as making sure that the DM appetite of the dairy cow is always satisfied.

The system involves feeding silage in conjunction with grazing grass. It is the best possible way of checking whether the grass supply is adequate. If the silage is eaten then more should be put out until cow appetite is filled.

Total Annual Requirement of Conserved Grass

The estimate of silage required through a winter was put at 9.6 tonnes/cow
Summer buffer feeding could be estimated as 2.0 tonnes/cow

Total 11.6 tonnes

As a very rough guide, it should always be the policy to produce 20 per cent more silage than is needed for the normal winter requirement. It is economic suicide not to have an adequate bulk feed supply for both summer and winter.

CHAPTER 9

HERD MANAGEMENT IN THE WINTER

We have seen how to plan for the winter feeding of the dairy herd and the scientific principles involved. The next step is to consider the day-to-day management of the herd.

From the examples of rations given in the last chapter it would logically follow that a herd has to be split into several groups in order to get the best out of the available foods. The requirements of a cow differ considerably at different stages of lactation.

The cow in early lactation requires a high-energy diet; little milk being taken from bulk feed. Mid-lactation cows can take some milk from bulk feeds, whereas late-lactation cows can take up to 12.5 litres from silage at the end of the winter.

FEEDING GROUPS IN THE DAIRY HERD

The need to split the dairy herd into feeding groups is likely to arise as the result of discussions concerning the general feeding of the herd and to make sure the individual cows receive the nutrition that they deserve with reference to their level of milk yield, stage of lactation and quality/availability of bulk feed.

The calving pattern in a herd is the most important factor which would lead to splitting the herd into feeding groups.

The two extremes of calving pattern are all the year round and block.

All the Year Round Calving

As milk production was originally closely related to the retailing of milk, and the housewife looked for a regular supply, production also had to be regular. Calving throughout the year has therefore been the traditional system practised on the majority of dairy farms.

Block Calving

This pattern of calving was introduced as a means of producing milk at certain times of the year when normal production was not able to ensure continuity of supply. As a result of general trends in milk production there evolved a considerable over production of spring milk (May–June) and a shortfall in production during mid–late summer.

Manipulation of the milk price (low prices in May–June to discourage spring milk and high prices in mid–late summer) has encouraged many farmers to move to block calving during the July to October period.

This policy results in a high proportion of the herd calving within a short period of weeks. From the feeding point of view it simplifies matters; all the newly calved cows are fed at the same level, since they will be at the same stage of lactation, giving similar quantities of milk and requiring the same ration.

Grouping of cows into separately housed groups is likely, therefore, to apply to all-year-round calving herds, where there are high yield and low yield producers together in the herd at the same time.

The necessity to form groups only applies to herds which have high yields; i.e. 6,000 litre average and higher. Dry cows should always be left in a separate group.

The normal grouping in a dairy herd would be for the following classes of stock:

(A) Dry cows and heifers within two months of calving The level of feeding for this category of stock should be designed to make the animals fit but not fat. This ties in with the idea that cows should be fit as they go dry and this condition must be maintained through the dry period.

Many years ago, cattle in this group would have received concentrates prior to calving. This technique has not been practised recently because of the realisation that good quality bulk feed can satisfy all the requirements of the dry cow. In fact there have been occasions when the silage has been too good for dry cows, tending to make them acquire too much condition before calving. Fat cows do not milk or breed very well and heavy calves at birth cause many calving problems.

(B) Newly calved cows and heifers—High yield The objectives for feeding this group would be to ensure that the group produce a peak yield which will result in the total herd production being achieved. Where high peak yields are necessary, this is likely to require a fairly high level of concentrates, perhaps up to 10 kg or more. It is at this point that grouping of cows becomes a real advantage. With the higher levels of concentrates being fed, it is necessary to split the total concentrates up into several feeds.

A general arrangement would be two parlour feeds of 4 kg and one out-of-parlour feed at 2 kg, or as required up to 4 kg per feed.

Dividing the feed up in this way, little and often, is a considerable aid to the more efficient use of feed. The size of this high yield group will vary since only the animals which will benefit from the extra feed allocation need to be in this group. Once the yield has dropped and there is no longer any need for the third feed, then cows can be transferred into the lower yield group.

(C) Mid/late lactation cows—Low yield Animals in this last group will be well into their lactation, served and probably PD'd (pregnancy diagnosed) in calf.

Extra feed is no longer required and the remainder of the concentrate ration can be steadily reduced until all the milk is produced from bulk feed. The key to feeding animals in this group is that they must produce milk well into late lactation using just best quality bulk feed fed ad lib.

In larger herds it is sometimes thought desirable to house the heifers in separate groups. Although every effort should be made to ensure that the heifers are well grown at the point of calving, it is possible that where there is limited access on a feed fence, or on an electric line across a self-feed silage pit, then heifers might be pushed out of the way or, in the case of the electrified wire, frightened from eating their entitlement of food thus restricting growth.

Grouping of cows also depends upon suitable buildings being available. Many of the traditional buildings used in dairying do not have the facility to be used in this way.

The biggest constraint on the ability of a cow to give higher yields of milk is likely to be feeding and management. Better feeding in early lactation will produce high yields. The other important factor to consider is that high yields should be obtained with the most economic usage of bought-in energy foods.

Cutting Feed Costs

By dividing the herd as suggested, the dairyman is given the opportunity to make the best use of home-produced bulk feeds. The general approach to feeding cows should be that those giving high yields are fed well and lower-yielding cows are made to produce on bulk feeds.

The essential management feature needed to arrive at and maintain a successful feeding policy is to record production on a regular weekly basis and attempt to correlate the milk yield with the food fed.

In early lactation, high yields are inevitably associated with a high level of feeding. The objective of the first one hundred days of lactation are that the cow should attain a good peak yield of between twenty-five and thirty litres. Body weight loss must be minimised to enable the cow to conceive to first service. Once she is in-calf, the milk yield tends to drop. The requirement then is to reduce the proportion of expensive bought-in food and increase that of relatively cheaper home-grown food, at the same time making sure that the milk yield does not fall at a rate faster than is normally acceptable. Dr Broster of NIRD has suggested that a fall of 2½ per cent per week or 10% per month is the normal expectation. With good-quality bulk feed the replacement rate of bulk for concentrates can be increased. If quality of bulk is poor then little replacement can be carried out.

The essential exercise is the matching of milk yield and feed input. This can only be achieved by regular monitoring of milk yield. In addition to showing clearly the value and effect of a feed programme, regular weekly milk recording also highlights problems which occur in the different groups of cows which calve each month. Once problems have been identified, they can be solved.

PRACTICAL FEEDING SYSTEMS

Dairy cow feeding has progressed a long way from the traditional feeding of cows tied by the neck for the whole winter in individual stalls. Virtually all dairy herds at the present time are loose housed. This development took place very rapidly as labour costs escalated during the fifties and sixties.

Straw yards were, and in fact still are, popular in the eastern counties, largely due to the availability of combined straw.

Unfortunately, the majority of dairy cows are kept in the western grassland areas of the UK where the supply of straw for bedding is limited. This problem was overcome by the introduction of cow cubicles where straw is used at a very low level. During the fifties there was an interest in forage farming, i.e. grazing of forage crops—kale, rape, etc.—during the winter months. As yields increased so it quickly became evident that the best policy for dairy cows is for them to be housed all through the winter. This change of view was greatly influenced by the increasing adoption of silage as the most important winter bulk feed for cows. The first widely adopted means of feeding silage to large number of cows was self feed.

This change of view was greatly influenced by increasing adoption of silage as the most important winter bulk feed for cows.

Self-feed Silage System

Self-feeding of silage offers considerable opportunities to cut labour costs, provided the layout is well planned and slurry disposal adequate. The system requires concentrating on silage-making to the extent of providing at least 1 tonne of silage per cow per month.

Depth of silage should not exceed 1.83 m and with 24-hour access, there should be 18–23 cm per cow of silage face, depending upon the breed of cow. If other bulk feeds are available then this may be slightly reduced. If cows are fed in batches, then 0.6–0.8 m of feed face should be allowed.

The simplest control for the cow at the face is an electrified wire.

The system does not allow individual silage feeding. It is also possible for timid cows and possibly heifers to be subjected to bullying.

Self-feeding is a solution to the problem of reducing labour cost, without at the same time reducing yield per cow. The basis of the system is a good-quality silage of about 23–25 per cent dry matter, with good digestibility and a uniform fermentation.

The facilities for storing silage vary greatly from farm to farm. Some farms use one large silage clamp for all the silage cuts. In other situations, separate buildings are used for each cut. Where just the single barn is available, a Dorset Wedge method may be used. In this way it is often the latest cut silage that is eaten first and the better first cut eaten in the later part of the winter. This fits in with providing quality silage to late-lactation cows at the end of the winter in order to produce milk from bulk.

The self-feed system is a group system of feeding, so individual supplementary feeding has to be carried out in the milking parlour. The amount of supplementary feeding varies according to the silage quality, the level of milk yield per cow and the stage of lactation.

Where the level of feeding has been improved, yields have consequently risen and in many cases it has been necessary to introduce a simple feed fence. This is a development which can be simply explained in the case of a cow giving 30 litres of milk and taking no milk from silage in early lactation (i.e. M + O).

The concentrates required at the rate of 0.45 kg/litre would be 13.5 kg. Up to 5 kg concentrates would be fed at each parlour milking, leaving 3.5 kg of concentrates to be fed down the feed fence in the middle of the day. This development also fits in with current research findings that concentrates should be fed 'little and often': perhaps two or even three times per day on the feed fence in addition to parlour feeding. The spread of feeding allows a better fermentation process to take place in the rumen; the pH is kept at a steady level instead of having violent fluctuations which tend to depress butterfat.

The concentrates on the feed fence can be of a cereal nature—barley, which is cheaper than compound cakes and also helps to utilise the high protein level available in good silage, over and above that required for a maintenance requirement. Sugar beet nuts are a very good concentrate for feed fence usage, and are suitable for the first 5–10 litres.

With high-yielding herds it may be necessary to feed for up to twenty litres on the feed fence; the last ten litres would need to be provided for in the form of high-energy compound concentrates.

The quality of silage, and the amount of supplementary feeding in the parlour and on a feed fence, are all factors involved in obtaining the optimal yield from any herd at the most favourable economic level. There is no definite answer, it is something to be sought for each year given our variable climate and its effect on quality and quantity of silage available.

Easy-Feed System

The easy-feed system of silage feeding is the logical development which farmers have to consider when the cow numbers in a herd grow beyond the level where the cows have the necessary width on the self-feed silage face. The system involves the provision of a feed fence, providing 0.8 m per cow. The silage is put out once per day with the use of a self-unloading trailer. The merit of the system is the ease with which any supplementary feed can be given. It is possible to provide self-locking yokes along the length of the feed fence where individual feeding is required outside the parlour.

It is desirable that the fence should be covered and illuminated otherwise the usefulness of the system cannot be fully exploited. There is the obvious requirement for additional building capital but this has to be weighed against the possibility of more cows being kept and the greater degree of control over the feeding of the groups of cows. The feed fence also gives greater flexibility in terms of feeding foods other than silage, such as hay, kale and potatoes. It is also particularly useful in the spring and autumn periods when fields are likely to be poached by grazing since grass can be brought into the system—a simple seasonal form of zero grazing.

The general system of zero grazing is a method whereby grass is cut all through the grazing season and brought to the cows by a form of self-unloading forage box. The system is expensive in equipment but saves fencing and laying on water to fields. Fields inaccessible, or too distant for grazing, can be put down to grass and intensively utilised. Production per hectare is claimed to be 7 to 10 per cent higher than in normal grazing systems; there is less waste of grass due to soiling. Criticism of the method arises from the high machinery costs involved and the large requirement for labour.

SUPPLEMENTARY FEEDING SYSTEMS

1. Feeding to Yield

The way in which cows should be fed has been a matter of much discussion over the years.

The system of feeding to yield has been well established for many years. The idea is that farm bulk feeds are used for maintenance and some production, with the concentrate providing the additional energy and protein for the higher yields sought after by many farmers. With the increasing quality of silage over the years, so the level of concentrate required per cow has reduced.

The short, sharp, shock of quotas forced farmers to realise that with only a limited production potential, their big concern for the future must be the improvement in the margin per litre. With this in mind, attention was increasingly given to the task of getting more milk from silage rather than from concentrates.

The traditional feed to yield system involved feeding 4 lbs to the gallon or 0.4 kg concentrates for every litre produced. The refinement of this system, coupled with the search for a better margin per litre has brought about a more specific definition of 'feeding concentrates to a level, as much or as little as is required, to produce the Total Herd Quota plus an appropriate threshold percentage.'

On an energy basis, concentrate is required to be fed at 0.45 kg per litre (see p. 138).

With bulk feeds being used for milk production as well, the actual kg/litre has changed.

e.g. A cow yields 25 litres of average quality milk, taking 7.5 litres from silage. The level of concentrate required would be

$25 - 7.5 = 17.5 \times 0.45$ kg/l

$= 7.87$ kg concentrates

The actual kg/l is therefore $7.87 \div 25 = 0.31$ kg/l.

At the stage in late lactation when a cow is taking 15 litres of milk from silage, the kg/l of concentrates required would be nil.

The modern approach of 'feed to yield' is applicable not only to winter feeding, but also for feeding during the rest of the year—see the section 'Summer herd management'. The system is capable of giving the best possible results and involves recording 'cause and effect' and acting positively on the results.

Feeding to yield in practice As a general guide, rates of feeding depend upon silage quality (to be checked by silage analysis), the quantity eaten and the yield of milk as shown in Table 9.1.

The table is a very practical rule of thumb system—the kg/litre used here is 0.4 kg/litre (lower than the scientific level of 0.45). If the level suggested by the table is used and is successful, then it shows that the silage is better than expected. If expected yield does not materialise then feeding should be increased by 1 kg, to find out whether this produces the expected yield.

As can be seen from the table, given a silage of an established quality by analysis, the level of supplementary feeding is varied according to the amount of milk given. When the appropriate level of concentrates has been fed, the cows on ad lib silage have the facility to satisfy their appetites on the silage face. If silage of the same quality is used throughout the winter, then as the milk yield decreases so the supplementary feeds can be curtailed. If the silage face improves in quality, as in the Dorset Wedge system where the cows move from third to second to first cut, this improvement should be turned into more milk from silage and again supplementary feeding can be reduced.

The table can be used where cows are at grass. For example, where a December-calved cow was giving 20 litres of milk and was taking 15 litres from grass, she would require a supplementary feed of 2 kg of concentrate food; this could be wholly cereal or sugar beet pulp nuts—as a rough guide, 1 kg = 2.5 litres of milk.

Table 9.1 Rates of Feeding

Milk from bulk feed	*Milk yield per cow per day (litres)* 10	15	20	25	30
	Concentrates to be fed per day (kg)				
M + 0.0	4	6	8	10	12
M + 2.5	3	5	7	9	11
M + 5.0	2	4	6	8	10
M + 7.5	1	3	5	7	9
M + 10.0	—	2	4	6	8
M + 12.5	—	1	3	5	7
M + 15.0	—	—	2	4	6

2. Complete Feeding

The system of complete feeding has been used in the United States and Israel for many years. In the United Kingdom, the system was introduced in the late seventies. The system is simply an extension of the idea of feeding a high-energy food little and often as in the case with herds on easy-feed silage.

The concept of a total diet is to give the cows a complete mix of available foods—silage, cereals, brewers' grains, straw and concentrates—which will provide for the total requirements of the cow in her expected dry matter intake. The total energy within the mix 'M' (i.e. total MJ) compared to the kg of DM matter intake 'D' gives the M/D or the energy concentration of the mixture. This is required to be between 8 and 12 MJ/kg according to the stage of lactation.

The density also has to be related to the level of milk yield. The mix should have a DCP of 11–12 per cent and a minimum fibre level of 14 per cent.

The system can eliminate parlour feeding and therefore allows milking to be carried out without reference to feeding operation.

The mixer wagons used in the systems can be fitted with load cells which allows accurate weighing of all feed on a group basis, but not to individual cows; the groups of cows are fed once per day. On the large arable farm which has many by-products, the system offers scope for feed economies. However, these tend to be balanced by the increasing level of machinery running costs and depreciation.

The system seems to offer considerable potential for increasing cow yields and compositional quality of milk, but to be financially successful the energy, dry-matter intake, protein and fibre levels need to be very carefully monitored.

Readers who wish to have more detailed information should consult *Cattle Feeding* by Professor John Owen, published by Farming Press Books.

3. Flat-Rate Feeding

The system of flat-rate feeding was developed as a means of avoiding the need to ration cows individually. It is concerned with the allocation of supplementary compound feed through the winter.

The compound feed is offered at a single predetermined level to a whole herd or defined group of cows through an entire winter feeding period. Silage is fed on an ad lib basis as with the feed to yield system. Once cows are placed in a feeding group they are not moved into any other feeding group.

The daily allowance of concentrates is reduced or discontinued once the cows are out to grass.

The idea of the flat-rate feeding system assumes that when the specified amount of concentrates is fed, any further additional nutrients will be met by the cows individual ability to fill its DM intake. The essence of the system is its simplicity.

Only farms which used parlour feeding are likely to save labour by adoption of this system since identification of cows for feeding is almost eliminated.

Some time is also saved since there is no need to calculate adjustments to rations. Successful operation of a flat-rate system without grouping demands a tight calving pattern. Herd replacement rates may be initially higher than normal.

Cows must be in good condition at calving, since there is a greater than usual possibility of underfeeding in early lactation. This may adversely affect fertility and increases the possibility of production diseases such as acetonaemia and milk fever. Results suggest that lactations of up to 6,500 litres have been achieved on the flat-rate system.

Table 9.2 shows a simple breakdown of likely lactation yield for four different calving seasons and at three different levels of flat-rate compound feed.

Table 9.2 Anticipated Lactation Yield (kg) Related to Month of Calving

Level of compound feed (kg/cow/day)	*Sept/Oct*	*Nov/Dec*	*Jan/Feb*	*Mar/Apr*
5	—	—	—	4500
7	5500	5250	5000	—
8	6500	5900	6000	—

Source: Ministry of Agriculture Fisheries & Food (2312) 1983.

Flat-rate feeding requires a plentiful supply of silage through the winter, approximately 10 tonnes per cow. This, together with summer silage requirements would total approximately 12 tonnes per cow per year.

CHAPTER 10

HERD MANAGEMENT IN THE SUMMER

Well-managed grassland is the cheapest source of cattle food any dairy farmer can have. At the same time, good grassland farming is one of the best ways of maintaining and enhancing soil fertility.

Therefore, on every dairy farm, grass should be regarded as the most important crop. Properly managed, it can give a far greater cash return per hectare than many cash crops.

Before outlining the details of successful grass management, three broad problems need stating.

Firstly, grass is seasonal in growth with a spring flush in May and a less marked flush in autumn; weather conditions influence the extent of the flush and the period over which it is maintained.

Secondly, there are marked changes in the feeding value of grass during its growing season. It can vary from being equivalent to a form of concentrate to being no better than straw.

Thirdly, when grass is at its best from a nutritional point of view it is highly succulent and laxative and raises certain difficulties in feeding.

The problem of seasonal growth is one that every grassland farmer must study if he wishes to have good-quality grazing over as long a period as possible and so cheapen his feeding costs tremendously.

A considerable extension of the grazing season is possible and a more even growth of grass can be maintained throughout the season by proper use of different varieties and strains of grasses and clovers, and by correct fertiliser application at the appropriate times.

The guiding principles are to cut or graze grass before it passes its best feeding stage, to give it adequate rest to recover, and to maintain soil fertility by adequate applications of potash, phosphate and nitrogen at times when the root systems are best able to deal with them.

A dairy farmer who sets out to establish good grass on his farm and to evolve a system of management to suit conditions and stock will be well repaid. He will be cheapening the cost of milk production and, when good leys are ploughed-in during a rotation, he will benefit from heavier yields of crops due to the improved fertility.

But it is just as important to make proper use of the grass as it is to plan to grow it wisely. This, too, is a subject that needs mastering if grass is to be exploited to the fullest extent.

The seasonal variation in feeding value has already been noted. Young, rapidly growing pasture has a dry-matter analysis value similar to a concentrate. Nevertheless, as a sole food for dairy cows it is deficient in fibre and too laxative. Thus the digestion of young grass is relatively inefficient.

FALL IN FEEDING VALUE

At later stages of growth there is a steady fall in feeding value and an increase in fibre content. Consequently, at the pre-flowering stage, grass has a dry-matter analysis similar to wheat feed middlings and the fibre content is at its optimum for efficient ruminant digestion.

Growth beyond the flowering stage to over-ripeness leads to a further fall in feeding value. Indeed, when much of our grass is cut for hay its feeding value, on a dry-matter basis, is less than half its value when young. Likewise the content of indigestible fibre has increased more than six times.

The level of soil fertility and the balance between grasses and legumes in the sward affects the feeding value at all these stages.

The assessment of the value of a pasture at any time needs considerable skill. Furthermore, the physical effort required of the cow in grazing must not be overlooked.

FERTILISER APPLICATION TO GRASSLAND

Modern fertiliser practice is related to the intensity of grazing, i.e. stocking rate, and the recommended rates of application per hectare. The amount of rainfall has a marked effect upon grass growth, and fertiliser applications cannot be effectively used unless there is adequate moisture in the soil.

With a good soil it is possible to stock dairy cows (500–550 kg

Table 10.1 Effect of Grazing System on Grass Quality

Pasture grass	*Dry matter content* %	*ME* *MJ/kg*	*DCP* %	*D value* %	*Fibre* %
Set stocking—close grazing	20	12.1	22.5	75	13.0
Rotational grazing—3-week interval	20	12.1	18.5	75	15.5
Extensive grazing—free growth after June	20	9.7	10.1	63	22.0

Source: MAFF, *Composition and Nutrient Value of Selected Feeds*, P2087, 1988.

The range of grass feeding is shown; the changes illustrated greatly affect the possible production from grass during the season.

liveweight) at a level of 2.2 to 2.5 cows per hectare (i.e. 0.45 to 0.4 hectares per cow). To achieve this it is necessary to provide fertiliser at the level of 300–350 kg N per hectare, 40–60 kg P_2O_5 and 40–60 kg K_2O.

The actual level needs to be closely related to a soil analysis which can be carried out by ADAS or commercial firms.

The nitrogen in the spring is applied as a straight dressing, whereas phosphate and potash are applied with nitrogen during mid-season, avoiding the risk of grass staggers. This is followed by straight nitrogen if required in the late August period.

The application of fertiliser has to be planned to correspond with the feed requirements of the dairy herd for fresh grass and grass for conservation. As fertiliser prices rise so it becomes important that all the grass grown is effectively used.

Grass surplus to grazing requirement should be cut and conserved. Ensiled grass does not deteriorate with age provided the clamp or silo is properly sealed.

To utilise fertiliser to the maximum, it may be thought necessary to install irrigation. This is a capital investment which has to be considered carefully. Less intensive stock rates (1.4–2.0 cows per hectare) require much less fertiliser, i.e. 200–250 kg nitrogen per hectare, and 25–30 kg of P_2O_5 and K_2O. At this stocking rate clover can play a useful part in the provision of nitrogen in the sward. Production from this type of sward is low and it is much more suited to extensive grazing areas, as found in marginal areas, with thin soils.

APPLICATION OF SLURRY

Slurry varies widely in manurial value. An average analysis would be 35 kg N, 11 kg P_2O_5, 58 kg K_2O in 10 m^3 of slurry. The use of slurry tends to build up potash-rich swards. This could lead to problems of hypomagnesaemia when the sward is grazed.

Where slurry is used, fertiliser dressings can be reduced. If slurry is applied to spring-sown crops in the spring, then due account can be taken of the N, P and K levels.

In the autumn, the value of nitrogen in slurry will be much less, since much will be lost by leaching over the winter. During the early nineties the EC is likely to bring out regulations concerning the handling of slurry.

SOILING OF PASTURES

This is a problem with heavy stocking. Herbage soiled by urine is not eaten by stock for 3–4 weeks, but herbage soiled by dung is unpalatable for much longer. Alternating grazing with cutting helps to maintain sward palatability, whereas zero-grazing offers an opportunity to intensify stocking rates beyond that obtainable by grazing only, and is a distinct possibility where land is highly valuable or liable to poaching by grazing stock.

CONTROLLED GRAZING

The objective in controlled grazing is to raise milk output per hectare by consumption of the grass at its optimum stage of growth, with the minimum of waste. Ideally, grass should be eaten off quickly and then given a rest period of varying duration during the growing season. In practice, there are three systems which are currently used with grazed grass:

1. Strip Grazing

This type of controlled grazing was introduced into the United Kingdom as the first attempt at controlled grazing. In this system a limited grazing area is allowed each day behind an electric fence. The fence can be moved once or twice daily. The system is a most efficient means of rationing grass as the grazed area can be very strictly controlled and the grass can be eaten off as closely as desired. Since the grass is used efficiently, it releases other land for conservation purposes.

The area of grass on offer each day must however be closely related to the requirement of the cows and matched to the availability of the grass. If the area is too large then utilisation is poor and the cows select particular grasses. When the area is too small it is likely that the cows' ability to produce milk from grass is being restricted. Under wet soil conditions, poaching will probably occur along the line of the fence. Recovery is likely to be slow where the stocking is too intensive and the grass has been grazed too low.

If the progress across a field is slow, grass at the far end of the field may well become too mature; therefore it is advisable to cut for conservation to allow a regrowth for grazing later.

Something like 0.4 hectare per day would provide grass for forty cows. This figure is only a rough guide as there are so many variable as to the amount of grass, density and height, current rainfall pattern and the level of nitrogen applied.

The system has a high labour requirement and this has accounted for its general decline in popularity.

2. Paddock Grazing

This system was introduced as an alternative to strip grazing, giving a high degree of grass utilisation without a high labour requirement. The system is based on the rotational principle of graze and rest. After being grazed off, the grass is allowed to recover with a rest period of 21–28 days, before being grazed again. After each grazing, it is important that the paddock should be top-dressed with nitrogen to help the regrowth. The amount depends upon the rest between grazings, but is usually at a level of 5 kg/ha N per day during spring and early summer.

Ideally the paddocks are for one day's grazing, but this only applies for part of the season. In the mid May to early June flush it is quite likely that the paddocks would need to be subdivided, one paddock lasting two days. This intensive use of paddocks will allow grass on other paddocks to be conserved for silage in the early season.

A reasonable guide to stocking density would be one hundred

cows per hectare per day. This has to be in conjunction with adequate rainfall and nitrogen top dressing. Recovery after grazing also depends upon these two factors.

As the season progresses to July, it may be that the one-day paddock is not able to supply sufficient grass. There is a modification of the system whereby the herd is split into two groups, a leader-follower system, the newly calved cows taking the best grass, followed by the low-yielders and/or dry cows.

Soiling tends to build up over the season and as rejection of grass increases it often becomes necessary to top the pasture. New growth is then encouraged.

A paddock system should be adequately watered, troughs should be placed between paddocks and there should be easy access to them. Movement to and from the paddocks is greatly helped by provision of farm roads. Ideally, a hard road is required with good access to the paddocks.

Paddock grazing achieved widespread popularity. Most farmers appreciated the discipline of the system and also realised the advantage of its efficient grass use coupled with flexibility.

Table 10.2 Merits of Grazing Systems
(Scored 0–5 against qualities: 0=low, 5=high)

	Strip	*Paddock*	*Continuous grazing or set stocking*
Simplicity of management	1	3	3
Flexibility	5	3	0
Low labour	2	4	5
Low cost	4	3	5
Efficiency of utilisation	5	3	5
Palatability	4	3	5
Grass production	5	4	1
Correct quality	3	4	5
Correct quantity	3	4	3
Poaching	2	3	5

Source: MAFF. Booklet 2050.

3. Set Stocking

The modern system of set stocking developed as an attempt to combine the traditional set stocking with the tighter stocking rates of the paddock grazing system and thereby increase profitability. High levels of nitrogen are applied in paddock grazing. Cows are given free access to grazing over a set area, and the fencing is removed. Stocking rates are, if anything, slightly less intensive than for paddock systems, at an average of five cows per hectare for the grazing area. Nitrogen applications are in the range of 250–350 kg/hectare; the regular application of one dressing per month (at 60 kg/hectare) eases the management aspect of the grassland.

Grazing management is very simple; cows are turned out when the grass is very short and the normal winter rations are continued. As the grass grows so the other foods are reduced and then stopped. With all the set stocked area available, the cows do not poach the grass in wet weather. As the season progresses, the grazing area should be increased, starting at 0.16 hectare/cow and increasing to 0.22 hectare/cow by July/August. Silage aftermaths can be included in the system. There is a marked difference in the sward between the more open type to be found on the paddock system and the dense sward of the set system.

With the close grazing, clover is more likely to be present; the dense sward also discourages the ingress of weed grasses.

The main reason for the popularity of the set stocking system is the simplicity of management; a practical criticism is that it takes longer to collect the dairy herd from a large set stocked area at milking time. On many farms the area is divided into two, a day and a night pasture; this may fit certain farm layouts.

The milk production on each of the systems mentioned is very similar. The reasons for a farmer adopting any one of the systems is more likely to be related to layout of the farm, soil type and the personal preference of the farmer.

VALUE OF WILTING

Ensuring that cows obtain highly nutritive grass and, at the same time, eat enough fibre and dry matter is a real problem.

Young, dense swards have a high moisture content. The water in and on the grass may approach 90 per cent of its weight, and the dry-matter consumption of a cow grazing it will be less than if she eats air-dry foods. In May grass DM can be as low as 13.5 per cent.

For this reason the late Professor Boutflour suggested mowing such grass and allowing it to wilt before grazing. This practice increases the dry-matter consumption of the cows; it is also worth-while as a preventive of bloat. It has been used successfully with high-yielding cows.

MANAGEMENT AT 'TURN-OUT'

It can be realised, therefore, that supplementary feeding of protein, under spring and early summer grazing conditions in this country, is largely unnecessary. The feeding of cows at grass—except during periods of drought—is largely a question of supplying supplements of high-energy value.

To counteract the laxative nature of young grass and to reduce the risk of bloat, the changeover from winter feed to summer grazing should be made gradually. In the earliest stages of growth—probably from early to late April—the cows should continue to receive a feed of silage (or hay) before they go out to grass in the morning.

An electric fence enables the grass to be rationed; where strip-grazing is not used, the cows should be allowed up to one and a half hours' grazing after the evening milking and then returned to a bare lie-back field for the night.

Once the grass has lost its laxative character, the feeding of roughage can cease, but it can be used as a measure of the grass supply by indicating whether the cows have sufficient dry matter available. See buffer feeding (p. 169).

By the time the grass reached the pre-flowering stage (20–25 cm long) most cows can be allowed to consume grass to appetite. Those yielding over 22 litres should continue to be rationed with grass and they should receive some supplementary feeding of concentrated foods.

It is instructive to point out, at this stage, that the extent of the rise in milk yield experienced when the cows first go out to grass is an indication of the low quality of the home-grown foods fed in late winter.

Under good management, where high-quality home-grown foods have been available in adequate supply, the rise should not exceed 10 per cent in total milk yield. If the rise is greater than this in spite of good management, then better quality home-grown foods are necessary.

Guide to Spring/Summer Feeding

Feeding cows at grass during the spring and summer poses many more problems than are ever encountered when feeding cows on silage during the winter.

Feeding on silage out of a clamp during the winter can be based on the silage analysis taken through the clamp and at the face. This information allows the rations to be calculated scientifically. Although silage cuts through the clamp may be different they do not vary on a day to day basis. This however is the first problem encountered when assessing grazing grass. With good warm conditions and adequate soil moisture, grass grows quickly. The outward signs of the first appearance of seed heads indicate lowered energy and protein coupled with increasing fibre. The possibility of continuous wet weather is a further complication.

Normal grass dry matter is approximately 20 per cent but in time of heavy rain the dry matter can be as low as 15–16 per cent. Such a low dry matter content means that much more wet grass has to be eaten to satisfy the full dry matter appetite.

With an all grass diet, 90 kg of grass at 20 per cent moisture would contain the necessary 18 kg of dry matter. On the other hand if the grass is only 15 per cent dry matter, then 120 kg must be consumed to provide the same amount of dry matter. The necessary intake of an additional 30 kg of wet grass would obviously causes problems in the digestion of this extra material. In the case of the newly calved high yielding cow, the extra grass intake would limit the cow's appetite for the concentrates needed.

Supplementing Late Lactation Cows at Grass

In May, grass has a tremendous potential for milk production.

The following example shows the situation with reference to an autumn-calving cow.

e.g. A 625 kg Holstein/Friesian cow taking in
19 kg DM of May grass at 12.1 MJ/kg DM
would receive 19 × 12.1 = 230 MJ
Less maintenance and grazing allowance during
late pregnancy = 88 MJ

142 MJ

With 4 per cent BF milk @ 5.4 MJ/litre this would provide for $\frac{142}{5.4} = 26$ litres.

This supports the idea that grass in May can support up to

26 litres of milk. Mention must be briefly made of the possibility of grass staggers in the spring so a magnesium supplement is usually fed.

Production from grass does not remain constant. Spring turns into summer and the quality of grass deteriorates as it is grazed for the second and third time. If there is a shortage of rain, then this not only encourages grasses to go to head but may also prevent cows from filling their total DM appetite.

By mid-summer, late July, the milk from grass situation has completely changed.

e.g. A Holstein/Friesian cow taking in some 70 per cent of the DM appetite made up of mid-season grass of 9.7 MJ/kg DM would receive:

Grass intake = 19 kg DM × 70%	=	13.3 kg
13.3 kg DM @ 9.7 MJ/kg DM	=	129 MJ
Less maintenance and grazing energy requirement	=	88 MJ
		41 MJ

With 4 per cent BF milk at 5.4 MJ/litre this would provide for $\frac{41}{5.4} = 7.5$ litres.

This highlights the potential problems which affect milk from grass throughout the summer.

When management fails to monitor production and allows it to deteriorate in this way, margins from grass can easily disappear.

Once it is realised that quality and quantity of grass are inadequate, then steps should be taken to remedy the situation. In this example, with an autumn-calved cow, there would be no attempt to feed concentrates. Generally silage would be supplied as a supplement. If this was given to make up the appetite then the calculated energy deficit would also be made up. There would be no expectation of a high milk yield with a late-lactation cow but if the energy deficit were not made up, then the animal would turn body weight into milk, losing condition and therefore finishing the lactation in a poor body state. This situation would have to be remedied before the commencement of the next lactation. The supplement to the ration could alternatively come from a previously planned availability of silage aftermaths, spring grass reseeds, or green forage, stubble turnips, rape, etc.

If farm resources are not available then extra supplement e.g. brewers' grains or stock potatoes, may have to be bought in. By arranging to have stocks of farm bulk feed available, much expense of this nature could be spared.

All the supplements mentioned would need to be rationed on a per head per day basis for as long as the grass shortage existed.

Since supplementing cows with other bulk feed at grass can be helpful, it is important to conserve all surplus grass in the form of third and fourth cuts of silage. A surplus of conserved grass will always be useful in a hot summer.

The only way in which a dairy cow can tell the farmer that she is short of grass, is:

(i) by reducing milk yield. Losses of up to 3 litres per cow occur when grazing is limited in either quantity or quality. Deterioration in cow condition will eventually occur if the problem is not remedied.
(ii) by taking supplementary silage when it is offered.

If silage is on offer all through the grass season, it gives the cow an opportunity to show that her DM appetite is not being satisfied. It is the task of management to make sure that all the necessary farm resources are available to allow the cow to do the job which she is expected to do.

The system of supplementing grass to the dairy cow by bulk feed is now generally referred to as buffer feeding (see p. 166).

Supplementing the Newly Calved Cow at Grass

For many years it was a widely held opinion that summer grass would provide all that was needed for summer milk production. The introduction of seasonality payments to encourage summer-calving and a consequent increase in late summer milk production has completely changed the situation. Since the level of peak yield required is related to total lactation yield the degree of supplement needed must be geared to reaching the appropriate peak yields. Having considered the difficulties of producing milk from grass from late-lactation cows in a dry summer, it follows that even more thought must be given to supplying the needs for newly calved cows, who are required to produce peaks of 25–30 litres in July–August.

Buffer feeding of silage is definitely necessary but if the appetite is satisfied with silage of only average quality then it is vital to feed concentrates to provide enough energy for the necessary peak yields to be obtained. The condition of grass is so variable that it is difficult to make suggestions other than to record milk yield regularly, certainly every two weeks, in order to find out whether milk yield is proceeding along the forecasted line of production.

The following example shows how the balance between requirements and energy supplied is assessed.

e.g. Friesian/Holstein cow (625 kg liveweight) calving mid August required to peak at 28 litres.

Energy Requirements	*Energy (MJ)*
Maintenance	65
28 litres @ 5.4 MJ/litre	151
	216
Energy Supplied by Ration	
Concentrate fed assuming M+5 litres from grass	
28−5 = 23 litres @ 0.45 kg/litre = 10.35 kg	
10.35 kg concentrates (as fed) = 8.9 kg DM	
Concentrates supply 8.9 kg DM × 13.5 MJ/kg DM =	120
Total appetite = 19kg DM	
Grass must provide 19−8.9 = 10.1 kg DM	
10.1 kg DM @ 10 MJ/kg = 10.1 × 10 =	101
	221 MJ

This simple calculation shows a requirement of 216 MJ and an availability of 221 MJ per cow/day. The figures used are not suggested as being exact, but rather are an estimate. There are too many variables for complete accuracy including:

(i) Liveweight of the cow. Assumed to be an average of 625 kg.
(ii) Milk quality. It is hoped that approximately 4 per cent BF will be produced.
(iii) Amount of milk produced from grass. Drought or rainfall, in the prèceding week could alter the M+ figure by ±5 litres.
(iv) The appetite of the cow. Ministry of Agriculture publications admit that this is difficult to predict, with variations of up to 20 per cent being possible according to the stage of lactation and pregnancy, as well as forage characteristics, M/D of diet and its DM content.

(v) the actual value of the grass. This can vary almost on a day to day basis, and between different fields, paddocks, seeds mixtures, soil type, etc.

The only way to formulate any policy of how to feed down-calving cows in mid-August is to decide on milk production targets and then to monitor production to ensure that what is required to happen actually does happen.

Some thinking farmers have taken steps to virtually fully house down-calving cows by the end of August.

This makes for a long winter, but stabilises many of the variables which influence the daily ration.

THE VALUE OF AUTUMN GRASS

The milk-producing value of autumn grazing is below that of similar grass in the spring flush.

This lower feeding value is generally appreciated by stockmen, but the scientific explanation is less clear. It is likely that the higher fibre content of autumn grass, despite its good protein analysis, is one of the possible reasons, together with lower sugar content.

In practice, the production value of autumn grass is at least nine litres per cow per day less than similar grazing in spring. Failure to appreciate this fact will lead the dairyman to be too late in providing extra supplementary food for the cows and heifers, and is often the explanation of a catastrophic fall in the milk yield of cows in early winter, particularly after a flush of autumn grass.

It is fatally easy to over-estimate the value of autumn grass to in-calf heifers and heavy-milking cows and so to cause a drop in their production which cannot be regained.

There seems to be a definite advantage in maiden seeds for autumn grazing. They appear to have a higher milk-producing value than older leys or permanent pasture, and where it is possible to have a small acreage available each year in the course of the rotation, it will prove particularly valuable for this purpose.

Other supplementary grazing crops will also be useful at this time of year. Those usually grown are rape, kale and ryegrass. Assuming that the pasture being grazed provides for maintenance purposes, which is probably true in most parts of the country until late September, about 12.7 kg of one of these green crops will provide for 4.5 litres of milk.

With the objective of high peak yields from autumn calvers, combined with the possibility of cold, wet September/October periods, it became a common practise to bring all newly calved cows in by early to mid-September and to put them onto normal winter rations. Mid- to late-lactations cows can be left out on what, for them, is capable of producing 10–15 litres. Concentrates can then be reduced where possible allowing high margins to be achieved.

If more severe weather conditions apply then late-lactation cows should be brought in for their own comfort and to reduce the possibility of the land being poached. Dry cows can be left out until November but again care must be exercised to make sure that they are in good body condition.

The farmer should also be aware of the possibility that when dry cows are brought inside and given silage of too good a quality they may be inclined to put on too much condition. A mixture of silage and straw is often suitable to prevent this.

ELECTRIC FENCING GUIDE

Some practical points in the use of electric fences are worthy of emphasis.

1. Follow the maker's instructions carefully in erection. Ensure a tight wire, effectively insulated and free of any possible earthing by contact with the crop being folded off.
2. Run the fence so as to include a water trough, or access to the trough from the area being grazed. Particularly on rape or kale, cattle show a marked desire to drink before the appetite level of consumption is reached.
3. Allow sufficient frontage or feeding space per cow. Two metres of wire per cow is sufficient.
4. In grazing large fields of grass, where the time required to graze the whole area will exceed a week, it is advisable to use a back fence. This prevents defoliating the sward as it recovers from the previous grazing.

In this chapter, the importance of good management in the utilisation of the increased productivity from new leys has been stressed.

All the grazing systems listed in Table 10.2 are capable of comparable milk production. The choice will depend very much on the farm layout and attitude to grassland management of the operator. The table summarises the various merits.

RELATING GRASS TO MILK

The only way in which the profitable relationship between grass production and milk yields can be established is to record the output of milk and to relate it closely to the amount of grass available. When cows go out to spring grass after the acclimatisation period, it is likely that by mid-May the grass will supply most of what the cow needs in the way of energy and protein. Mention has been made of the possibility of expecting too much from the grass. However, problems tend to occur in late July when the grass supply becomes short in either quality, quantity or both. The question of supplementation then becomes a management question. If the yield starts to decline at a rate greater than normal, the question is what will it cost to reduce the rate of milk decline. The problem is solved by feeding supplements to one group of cows to see if the milk can be stabilised. A simple comparison of cost of feed against value of milk recovered will provide an answer. If twenty pence was required to regain milk to the value of twenty-five pence, the cost benefit is not too clear-cut. However, the other factor to be taken into account is the level of milk yield likely through the remainder of the grass season, with the possibility of an autumn flush of grass. A higher milk yield from supplemented cows would then be obtained wholly from bulk feed. This approach puts the issue of supplementary feeding into the category of a management decision, rather than one simply concerned with nutrition.

THE LARGE HERD AT GRASS

With the general increase in the size of herds, the question of farm layout to facilitate movement of cows to and from pasture requires careful consideration. Access to paddocks is best provided from a central 'race' with wide gates opening across the race to control the herd and reduce poaching.

Herds of over one hundred cows are difficult to accommodate as one herd under grazing conditions. The set stocking systems tend to deal with some of the problems such as poaching. On the paddock system the solution is to divide the herd in two groups, possibly high- and low-yield. The nutritional requirements of the groups are different; mention has been made of the leader-follower system which has been used successfully in many cases.

It is possible then for the large dairy herd to be more efficient at utilising grassland; it offers the possibility of grouping cows at grass, a system which would be impracticable with a small herd.

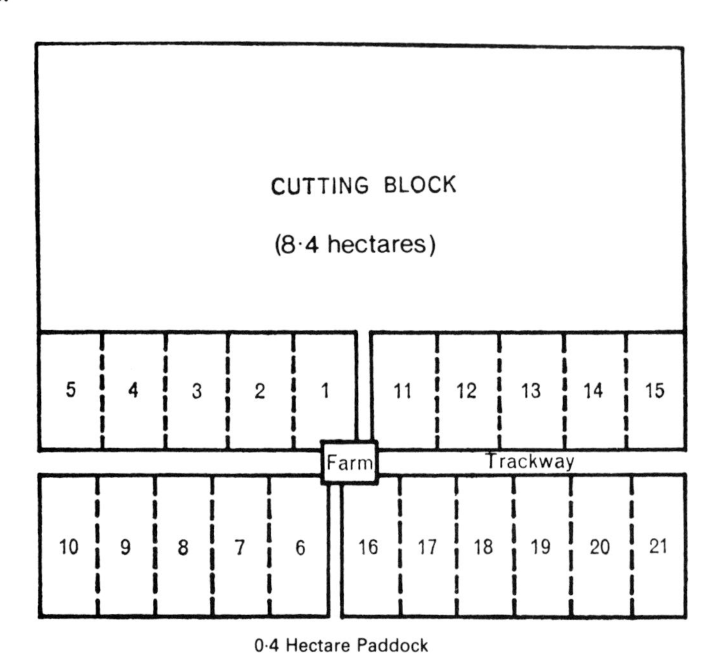

Figure 10.1 Paddock layout

CHAPTER 11

MODERN MILKING EQUIPMENT

In the last decade rapid developments have taken place in farm machinery, largely due to the adoption of the new technology of electronics. This is particularly true with the machinery used for the milking and management of dairy cows. It has enabled dairy farmers to adopt aspects of automation which relieve them of the more tedious and time consuming jobs, allowing them to concentrate on managing their cows to produce a healthy, wholesome and cost effective product.

However, we must not forget that the physiology of the cow and the basic principles of milk extraction by machine remain unaltered and a revision of these facts is a useful starting point for this chapter on modern milking equipment.

THE PHYSIOLOGY OF THE UDDER

The udder consists of four separate 'quarters' which are attached to the cow by a strong central ligament. Each quarter contains many milk manufacturing cells or, to give them their correct name, alveoli. The interior of the alveoli is lined with myo-epithelial cells which extract nutrients from the blood supply and convert them into milk. The milk is therefore held in the quarter rather like a sponge holds water.

When the cow is stimulated to 'let-down' her milk, a hormone called oxytocin is released from a gland near the brain, called the pituitary gland. The hormone is carried to the udder in the blood stream of the animal. Oxytocin causes the small muscles which surround the alveoli to contract, increasing the milk pressure in the milk cistern and ducts. If the muscle at the end of the teat is subjected to an air pressure which is below atmospheric air pressure, then the pressure differential between the end of the teat and the inside of the udder will be great enough to allow milk to flow.

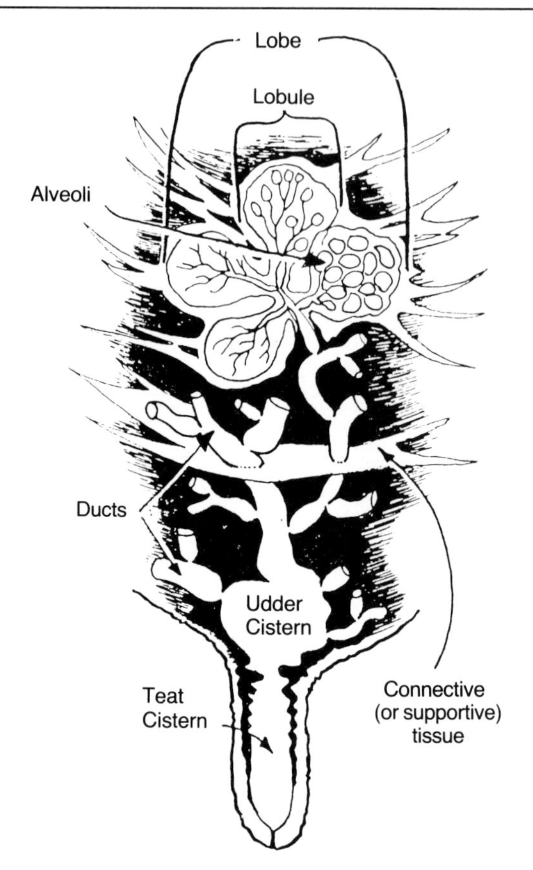

Figure 11.1 Cross-section of a cow's quarter
Harvesting Your Milk Crop, Turner

When the calf suckles the cow milk let-down occurs naturally. But with machine milking the stimulus is provided by elements of the milking routine. It is important that the animal co-operates with the milking machine, and the operator can assist here by applying the teat cups as quickly as possible after let-down has commenced, since the effect of the hormone lasts for only 6 to 8 minutes.

THE DOUBLE-ACTION TEAT CUP

Almost all milking machines use a teat cup which attempts to imitate the way in which a calf milks a cow. As the calf swallows

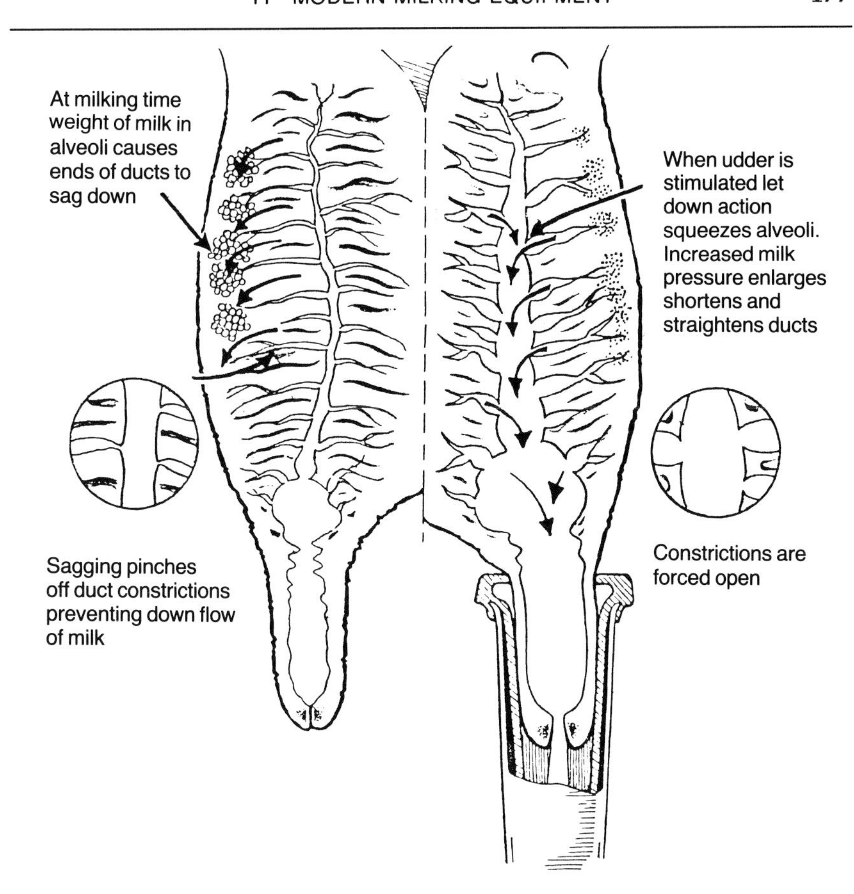

Figure 11.2 Cross-section of the udder
Harvesting Your Milk Crop, Turner

the milk, its mouth massages the end of the teat, promoting bloodflow away from the teat end. If this non-milking phase did not occur the accumulation of blood would cause discomfort to the cow and damage to the delicate teat end tissues.

The double-action teat cup reproduces the milking and non-milking phases by using a flexible synthetic rubber liner in a stainless steel or plastic teat cup (see Figure 11.3).

The liners are opened and closed by pressure changes between the liner and the teat cup, which are brought about by the operation of the pulsator. Depending upon the type of pulsation used, the four liners may open and close at the same time (simultaneous pulsation) or in pairs (alternate pulsation). The two phases of liner movement are shown in Figure 11.4.

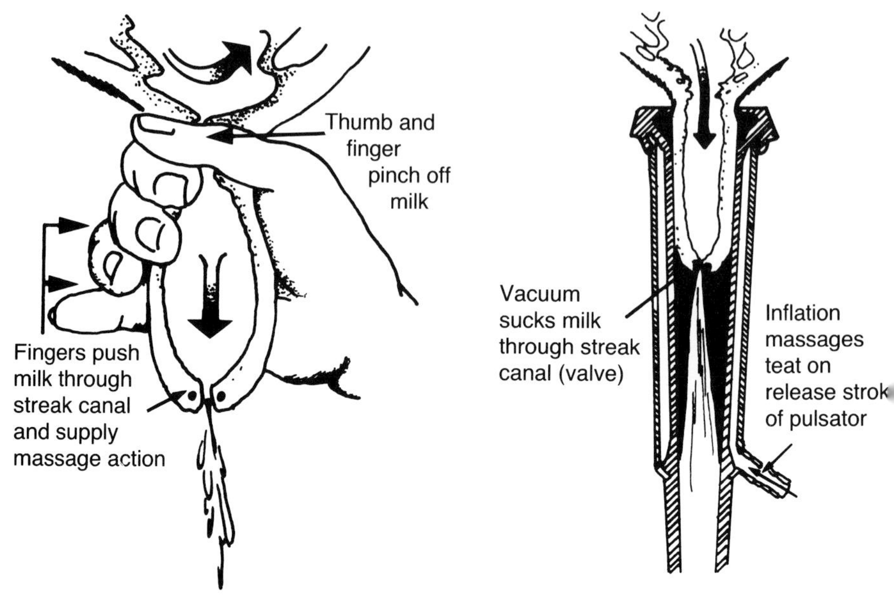

Milking a cow by hand uses one principle while the milking machine and the calf use another.

In hand milking you squeeze out a teat full at a time and massage the teat while doing it.

The machine and calf suck the milk out in intermittent strokes. Letting air into the shell periodically allows the inflation to collapse around the teat and to supply the massage action so necessary to avoid congestion of blood in the teat walls.

Figure 11.3 Milking by hand and machine
Harvesting Your Milk Crop, Turner

VACUUM AND THE VACUUM SYSTEM

The earth is surrounded by a layer of air which we call the 'atmosphere'. The air has weight and is kept close to the earth by the pull of the earth's gravity. The pressure of the air acts upon the surface of the earth and on all things on the surface: buildings, animals and human beings. In fact the total force exerted on a human body at sea level amounts to several tonnes. Atmospheric air pressure at sea level has a nominal value of 100 kiloPascals (kPa) or 1 bar (1,000 millibars) and an Imperial value of 14.50 lb/in^2. When the term 'vacuum' is applied to the milking machine, it means any pressure below that of the surrounding atmosphere. The working vacuum levels are generally decided by the type of milking equipment, i.e.

high level pipelines in a cowshed	)	
high level recorder jars	)	50 kPa
eye level recorder jars	)	
low level recorder jars	)	
low level pipelines	)	44 kPa
bucket plants	)	

In Imperial measurements these figures relate to 15 in. Hg. (inches of mercury) and 13 in. Hg. respectively. Small changes may be made to the vacuum levels quoted to cater for variations in plant design. A point worth remembering is that vacuum levels must be correct for milking and for circulation cleaning of the milking equipment.

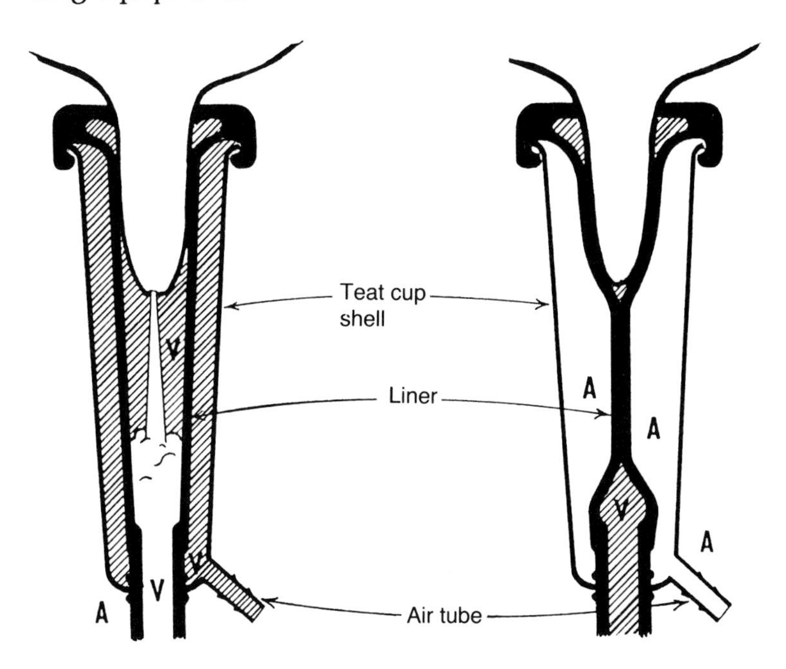

MILKING: Vacuum (V) applied to exterior of the teat opens teat canal so that milk is removed
REST: Periodic air (A) admission to the space between liner and shell permits closure of liner for massage

Figure 11.4 Teat cup action—milking phase and rest phase
Mastitis Management, W. Nelson Philpot

Figures 11.5 and 11.6 show the pipe and component layout of milking installations. The diagrams are taken from British standard (BS) 5545, which advises standards on the vocabulary, construction and testing of milking machines. The publication is available from HMSO.

THE VACUUM PUMP

All modern milking machines use a sliding vane type of vacuum pump as shown in Figure 11.7.

The pump consists of a heavy steel rotor, mounted off-centre in a cast iron housing with close-fitting end plates. Rectangular vanes made of synthetic material are a sliding fit in the radial slots of the rotor. As the rotor turns, the vanes are thrown outwards by centrifugal force, contact the smooth inner surface of the housing, and thereby create air chambers between adjacent vanes. The volume of the air chambers varies because of the eccentric position of the rotor and this is responsible for the pumping action. As an air chamber moves past the inlet port its volume increases. This causes a pressure drop and air flows into the chamber. When the chamber approaches the outlet port its volume decreases, causing an increase in pressure and the air is pushed out to the atmosphere. The rotor can be driven by several alternative types of prime mover: electric motor, pto shaft drive from a tractor, belt drive from an auxiliary engine or a hydraulic pump and motor set. The edge of the vane which is in contact with the honed inner surface of the rotor is lubricated and sealed by a thin film of low viscosity oil. It is most important that the correct grade and quantity of oil is introduced into the pump.

Pumps manufactured since the introduction of BS 5545 will have a specification plate on a prominant part of the pump displaying:

(i) air extraction capacity in l/min at 50 kPa
(ii) electric motor size in kW
(iii) oil consumption in cc's/hr
(iv) pump shaft speed in rpm

Exhaust back pressure tests have been introduced by the milking machine testing agencies and a reference figure for this test may soon appear on vacuum pump specification plates.

At an extra cost, at least one manufacturer will provide a version of a liquid ring pump to replace the conventional sliding

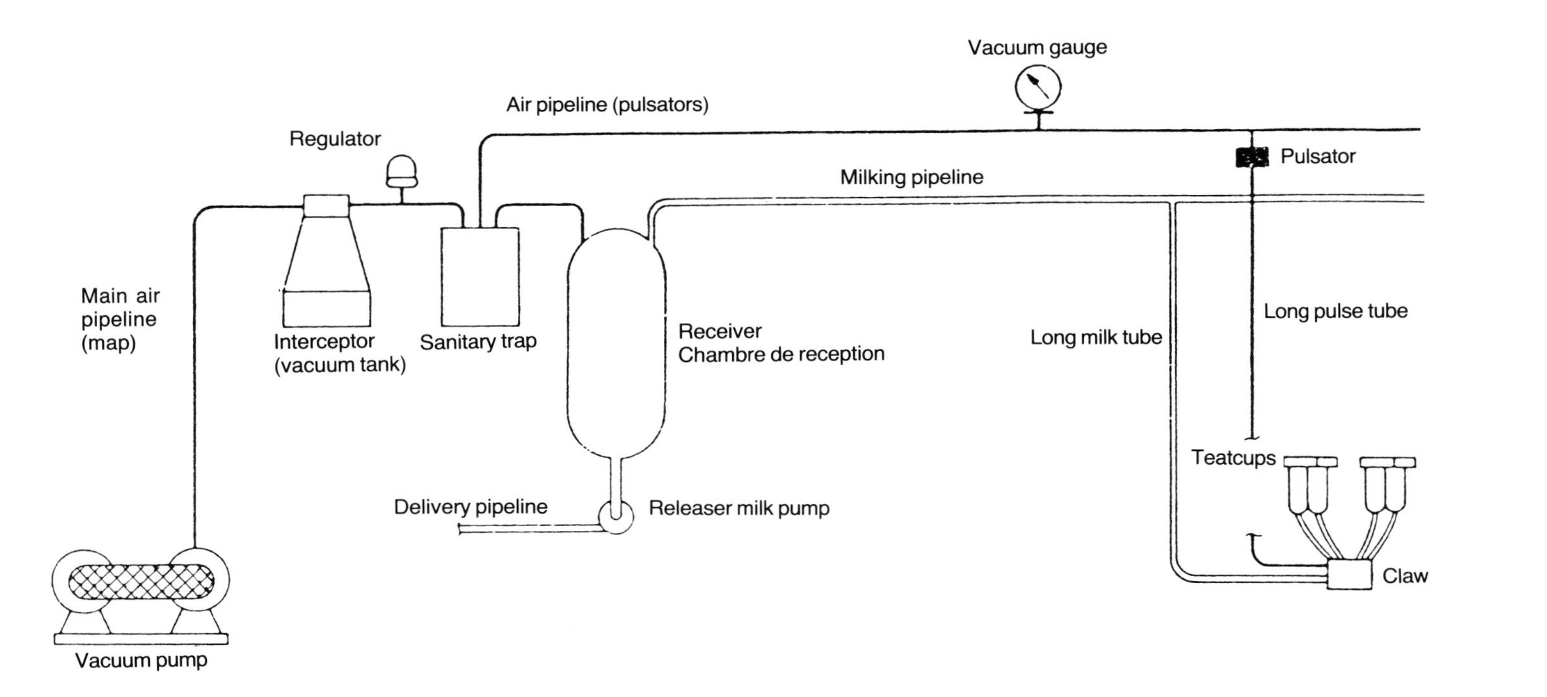

Figure 11.5 Milking pipeline machine
British Standards institution
Note: A mechanical releaser incorporating a milk receiving chamber may take the place of the receiver and releaser milk pump.

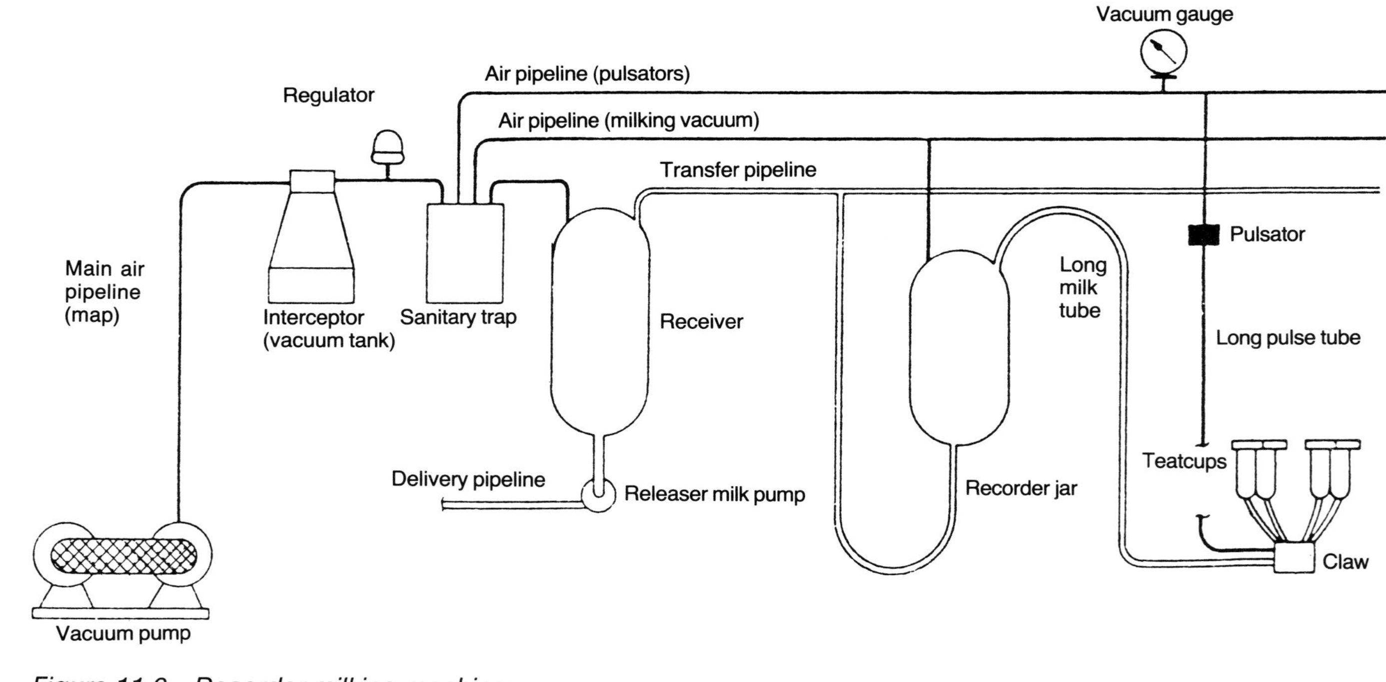

Figure 11.6 Recorder milking machine
British Standards Institution

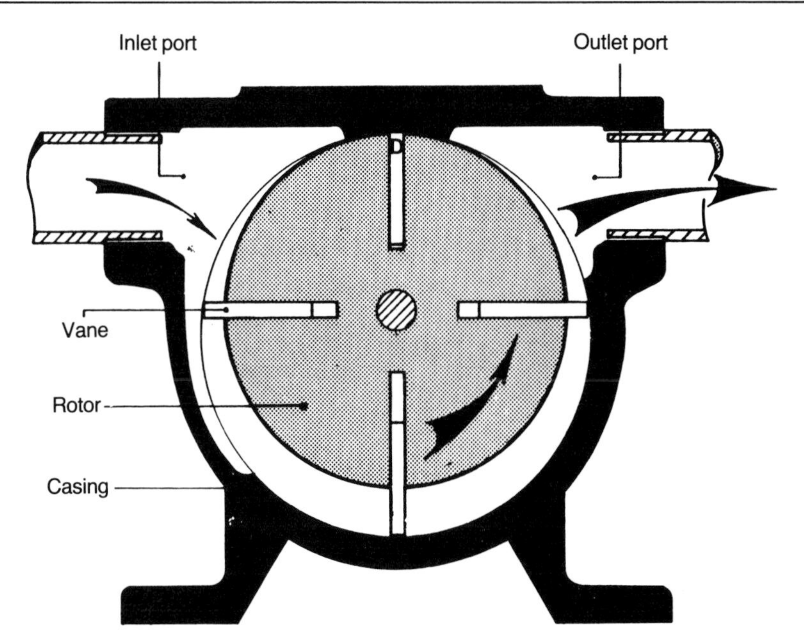

Figure 11.7 The vacuum pump
Machine Milking, Technical Bulletin No. 1, NIRD

vane vacuum pump. The liquid ring pump has been used in the food industry for many years and its introduction into agriculture to work as a vacuum pump is a new venture for the UK. A long working life and lack of conventional lubrication requirements may make its purchase a cost effective proposition for the larger milk producer.

Electronic devices in the form of liquid level controllers monitor pump oil present and provide visual and audible warnings of low oil level.

THE VACUUM REGULATOR

The function of the vacuum regulator is to maintain a constant vacuum level inside the milking machine. The speed and accuracy with which it performs this task is a measure of the regulator's sensitivity. There are several internal mechanisms which are utilised working on the principle of pressure difference. Figure 11.8 illustrates the weight operated vacuum regulator.

As the vacuum level rises, a point is reached where atmospheric air pressure in the air inlet overcomes the weight acting on the valve. This causes the valve to be lifted off its seat allowing air to flow into the main air pipeline (m.a.p.)

Vacuum pumps capable of extracting large quantities of air are used on milking installations where there is a proliferation of vacuum operated equipment, and consequently such pumps must be matched by more sophisticated and sensitive regulators. If a simple deadweight regulator were used where the vacuum level is sensed immediately underneath the valve which allows in the atmospheric air, then the large air flows and air turbulence would upset the regulator, resulting in unstable vacuum levels in the milking plant. To overcome this problem a remote sensing regulator is used. Figure 11.9 illustrates the remote sensing regulator. The vacuum level is sensed in the m.a.p. upstream towards the first milking unit.

From the sensor, the vacuum signal is transferred to the valve which will allow atmospheric air into the m.a.p. to give a stable working vacuum level.

Whichever type of regulator is used, the valve must be fitted between the interceptor and the milking plant (see Figures 11.5 and 11.6).

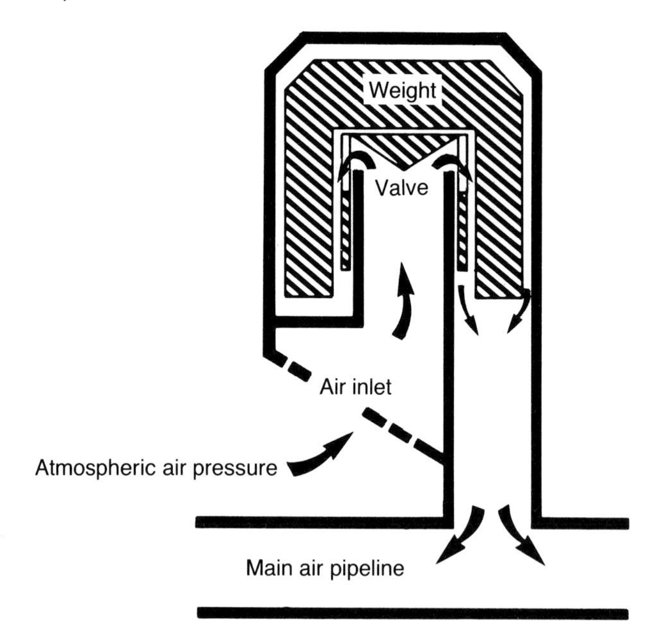

Figure 11.8 Weight operated vacuum regulator
Manus Technical Manual

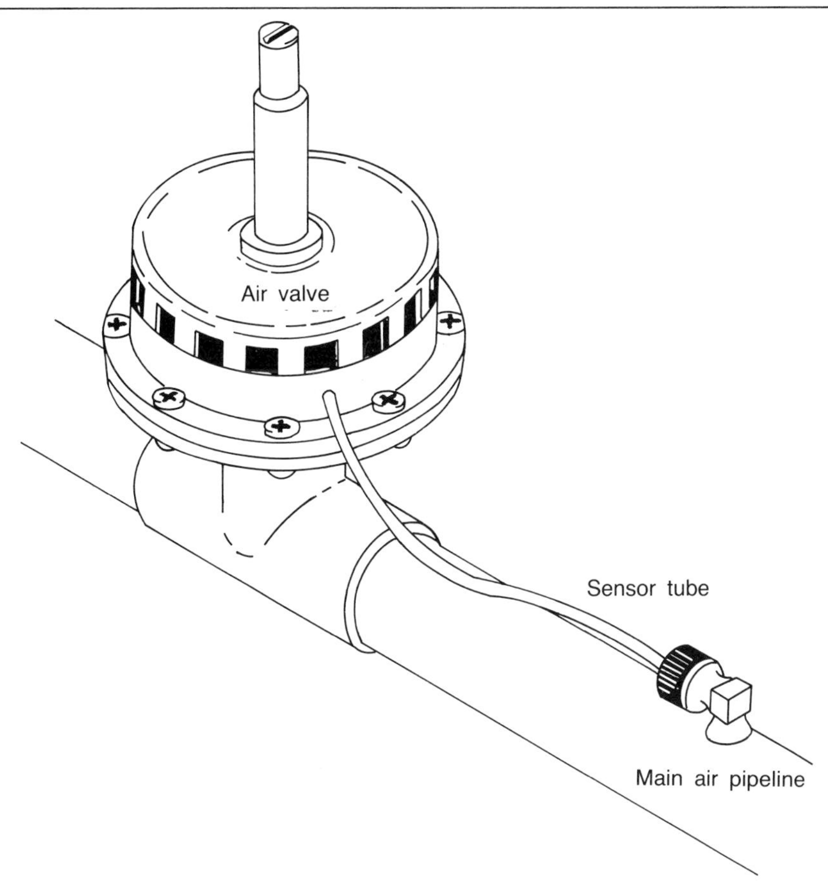

Figure 11.9 The remote sensing vacuum regulator
Fullwood Technical Manual

The demand for an adequate and stable vacuum supply has brought about radical changes in the design of the pipework in the main air pipeline and the air pipeline (pulsators). A large cylindrical plastic tank with a volume of 35 litres is situated between vacuum pump and parlour with interconnecting plastic pipework of 75 mm internal diameter. The smooth internal surface of the tank and pipeline aids air flow through the system. The tank also provides a suitable mounting point for the regulator valve.

Electronics have been used to monitor abnormal vacuum levels at one or more points in the milking plant and provide the operator with a visual and audible warning of dangerously high or low vacuum levels.

PULSATION

The liner of the double-action teat cup opens and closes during milking (see Figure 11.4) as a result of the pulsator action. The order in which the liners open and close can be described as

(i) alternate
(ii) simultaneous
(iii) sequential

An explanation of these terms would be:

(i) left side of the udder milking and the right side resting (one beat of pulsator) then, left side of udder resting and right side milking (one beat of pulsator) and so on. When put in number form, it is written 2 × 2.
(ii) all four liners open and close at the same time, written—4 × 0.
(iii) simultaneous electric relays which are switched on and off at staggered intervals (0.33 second), to provide an even vacuum demand.

There are two important definitions which relate to pulsation:

- Pulsation rate: the number of pulsation cycles per minute. Typically 50–60 per min.
- Pulsator ratio: the time which the liner is open (milking) compared to the time which the liner is closed (resting). This is expressed in milliseconds or as a percentage e.g. 70:30, 60:40, 50:50.

A wide range of mechanical, pneumatic and electrical mechanisms have been successfully used for operating pulsators. They can generally be classified as 'relays' or 'self-contained' types.

Relay pulsators are controlled by a central control box, which contains a transformer (to reduce 240 volt AC down to 12 or 24 volt DC) and the necessary electronic circuitry to provide the correct pulse timing and pulse duration. The control box sends a signal to each relay which, depending on its construction, may provide pulsation for one or two milking units. Figure 11.10 illustrates an electromagnetic relay pulsator giving simultaneous pulsation to one milking unit.

When two electromagnetic relays are fitted together in one housing they can be programmed to operate in several different ways (Figure 11.11).

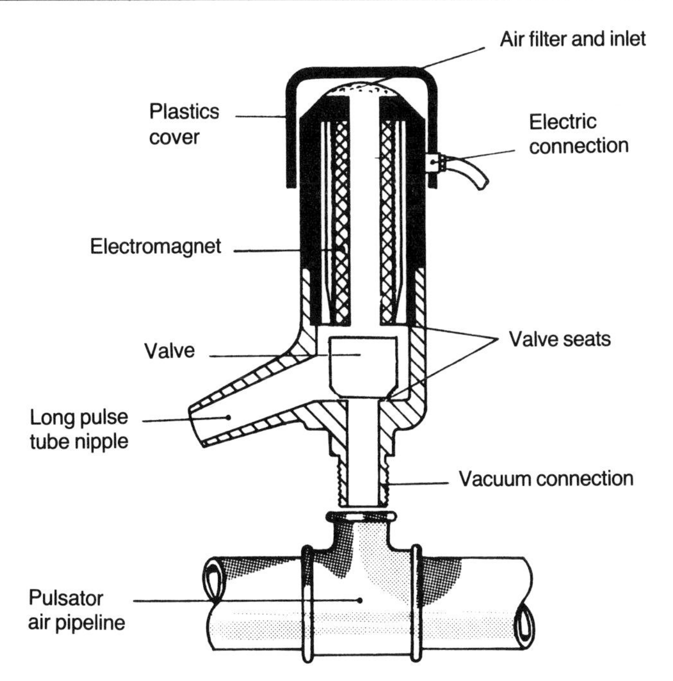

Figure 11.10 Section through the Fullwood electromagnetic relay pulsator
NIRD Technical Bulletin

Problems can occur if the central control box develops a fault, since this fault is passed on to all the relays and eventually to the cow in the form of incorrect liner movement, which may trigger off a case of mastitis.

Alternatively the relays can be driven by their own individual electronic controller, which only requires a low voltage power source. The dealer's technician can change switches on the relay's circuit board to alter pulse rate and ratio. This means that the pulsator can be matched to a suitable vacuum source and will be capable of providing pulsation for cows, sheep, goats, buffalos, camels etc.

If you wish to retain the existing simple relays, then it is possible to change pulse rate and ratio on special central control boxes, which once again will provide a very versatile pulsation system for many classes of livestock.

Adjustment of these individual relays or master control boxes should only be made by the manufacturer's agent or dealer,

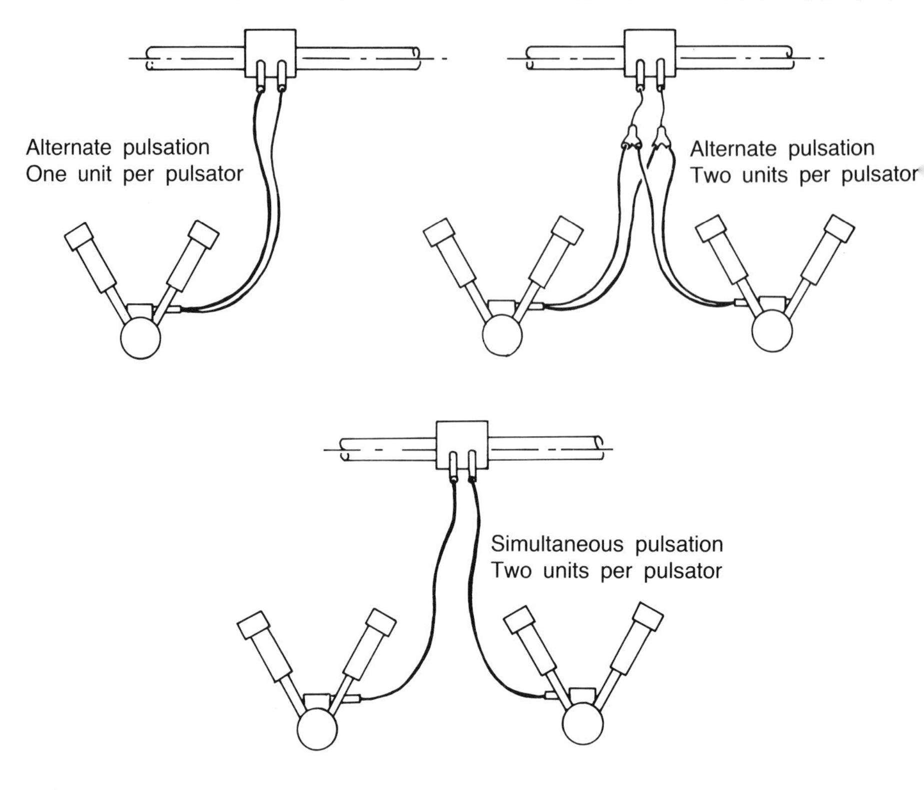

Figure 11.11 Electromagnetic pulsator used for various forms of pulsation
Gascoignes

who will have the necessary testing equipment to check the performance of the system after adjustment has taken place.

Pneumatic pulsators are invariably self-contained which makes them independent of a separate power source. Each milking point has it's own pulsator and timing mechanism. The photograph on page 189 shows a selection of pneumatically self-contained pulsators which provide 'alternate pulsation' to the teat cups.

Smoother milk flow and a more even vacuum demand are the claims of this type of pulsator. Many of these pulsators can be adjusted for speed and ratio, but once again must be checked for performance using specialised equipment.

One very important maintenance requirement of any pulsation system is the provision of a clean air supply when the milking machine liner is in the rest phase. To achieve this, many electric relays (simple or sophisticated) are fitted with a separate clean air supply pipeline, incorporating an air filter sock.

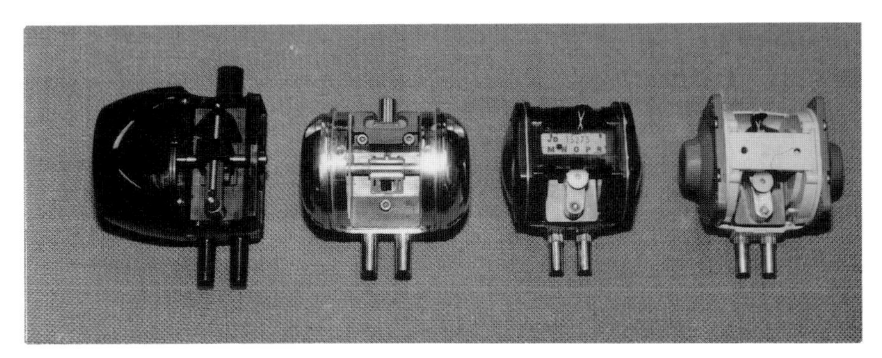

A selection of self-contained pulsators.
Cheshire College of Agriculture.

THE CLUSTER

The four teat cups and the claw, together with their connecting tubes are generally known as the cluster. The teat cup shell is usually made of stainless steel, although plastic disposable ones are available complete with bonded liner. The liner is made of synthetic rubber, which resists attack from milk fats and cleaning materials. There are many different sizes of liner to match the teats of the various breeds of cow.

The action of the opening and closing of the liner tends to cause a surging of milk from the claw, up the short milk tubes, to the teat ends. This 'teat impact' is thought to be responsible for cross-infection of the udder and considerable efforts have been made in liner and claw design to prevent impaction taking place. These innovations have included one way valves and plastic discs between the barrel of the liner and the short milk tube.

The claw provides the connection between the teat cups and the long milk and pulse tubes. The volume of the claw is very important, as any restriction in milk flow at this point would seriously slow down the milking process. The minimum volume of the claw bowl is 80 millilitres, but many exceed this being several hundred millilitres in volume. This provides a reservoir to hold the milk preventing it being washed back to the cow's teats.

As milk flows down the short milk tubes from the teat cups it passes into the claw, where an air admission hole introduces air

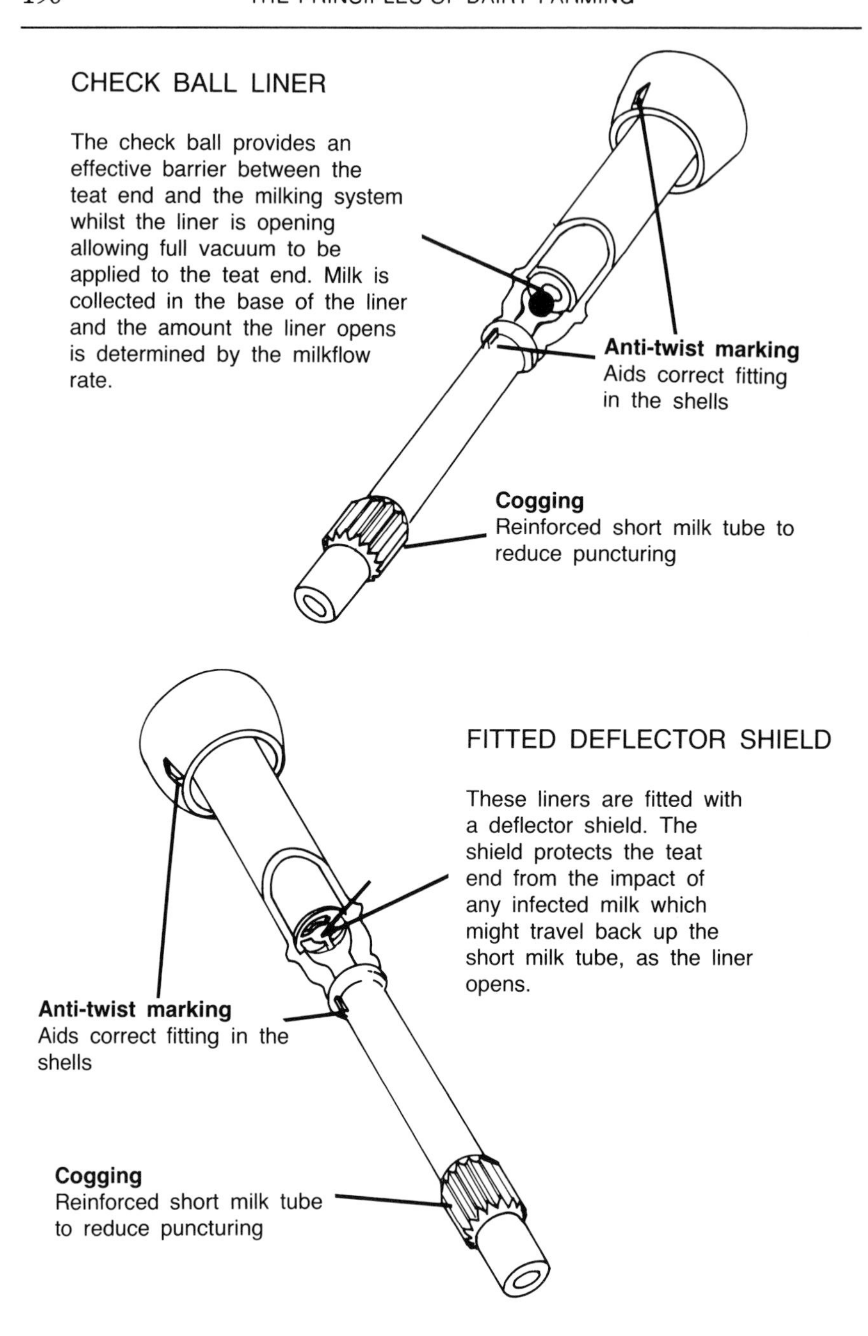

Figure 11.12 Check ball liner and deflector shield liner
Alfa Laval

at a controlled rate in order to help push the milk up the long milk tube. This air helps to reduce the weight of the milk in the long milk tube, thereby moving it quickly away from the claw. The volume of the air admitted can vary between 4 to 12 litres per minute. No attempt should be made to increase the size of the air admission hole, since excessive amounts of air will cause a breakdown of fats, and possibly lead to milk spoilage.

A different approach to cluster design is seen with 'hydraulic milking' in Figure 11.3.

The claw piece bowl contains stainless steel ball bearings within the upturned milk ports. When the liner is open (milk phase) gravity allows the ball bearings to drop, preventing milk flow

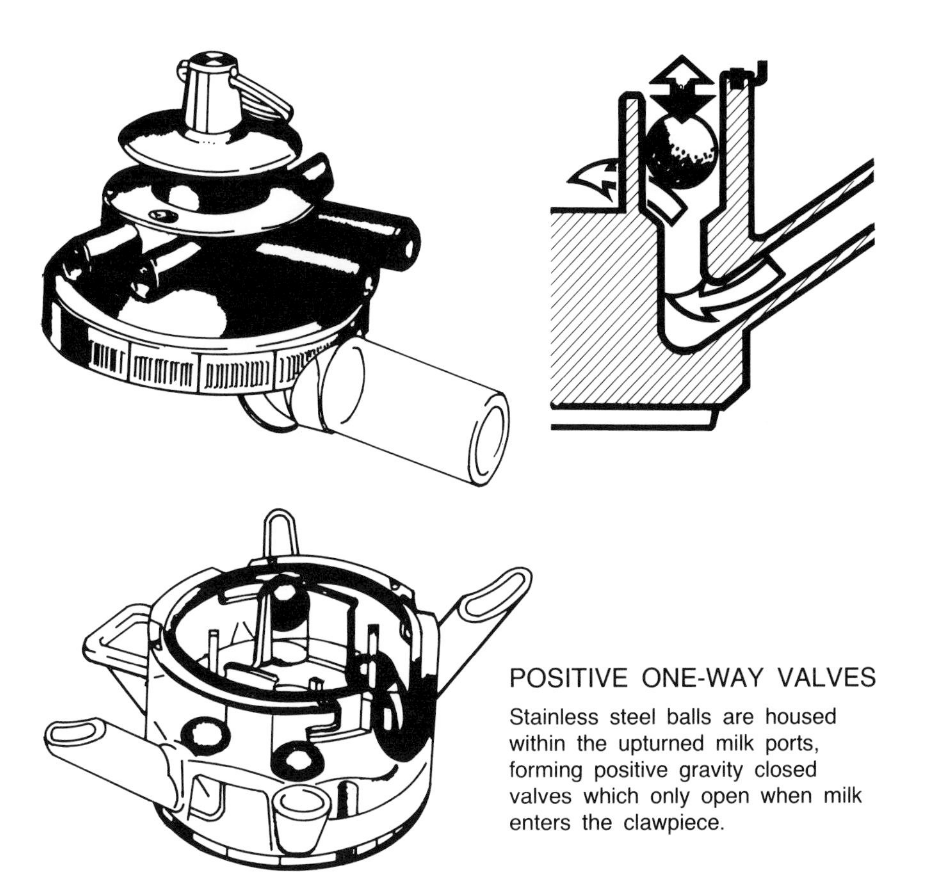

Figure 11.13 Hydraulic milking clawpiece
Ambic

into the claw. The action of the liner closing, going into the rest phase, forces milk down the short milk tube, unseating the ball valve and allowing milk flow into the claw. Claimed advantages are a more gentle milk action on the teats and the elimination of cross contamination and teat end impacts.

When removing the cluster from the cow it is useful to have a vacuum shut off clip on the long milk tube or an automatic shut off valve incorporated into the clawpiece bowl. This valve has another function, namely that it will close thereby preventing large quantities of atmospheric air entering the milking plant in the case of a milking unit being kicked off.

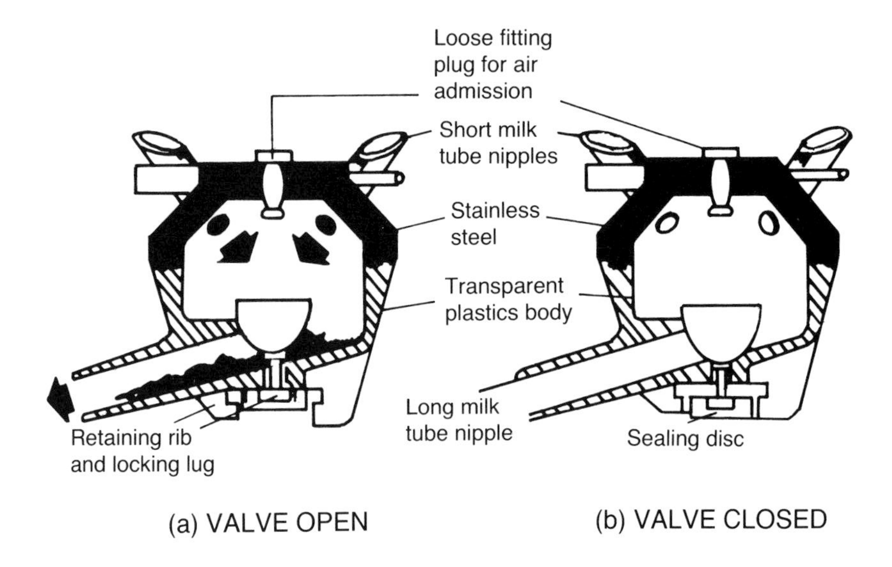

Diagram of the Alfa-Laval claw with automatic vacuum cut off. The valve can be locked in the open position for washing.

Figure 11.14 Automatic vacuum shut-off valve in clawpiece
NIRD Technical Bulletin

AUTOMATION FOR THE DAIRY HERD

Automation is playing a major role in several aspects of dairy farming apart from milking itself.

Automatic cow identification—The cow wears a collar or ankle strap, containing a small radio transmitter. Each transmitter is set at a different frequency. As the cow goes through an identification loop into the milking parlour or puts her head over a trough, an active coil starts her radio transmitter, the signals are then picked up by an identification processor. Once the cow has been identified, automatic milk recording and feeding become feasible.

Automatic milk recording—The performance of each cow can be checked by recording the amount of milk she gives during her lactation. Originally all recording was done manually, but this is time consuming and involves writing milk yields onto sheets in the parlour which is a messy business. Herds which subscribe to the National Milk Recording scheme must be recorded and sampled each month by NMR staff.

Automatic recording can take place every time milking is in progress by strain gauge beams which support recording jars, or, if milking is direct to pipeline, by milk meters. A digital read out at each milking point shows the amount of milk given. A useful addition to the system is a personal computer which will accept the recording via an electrical link between the milking installation and the farm office. All that the operator has to do is identify the cow in the stall before the milking unit is attached to the udder.

Feeding cows in and out of the milking parlour—There will always be the debate about whether cows should be fed in the milking parlour. It is not the purpose of this text to discuss such issues, but to summarise the options available.

With automatic cow identification, once the cow has been recognised at a certain stall, then the herd management system could feed her the correct amount to sustain a pre-determined level of milk production.

After manual identification of the cow, the operator could punch the number into a key pad on a feeder control box, which would then operate the feeder. In a much simpler parlour, say an abreast type, moving a lever which is directly connected to the feeder, will dispense a pre-set volume of feed.

If the feeder can weigh the concentrates rather than measure them by volume, then this should give greater accuracy. One method is for the mechanism to sweep feed into a balanced tray, which tips over at a pre-determined weight, allowing the feed to fall into the trough.

The major advantage of the out-of-parlour feeder is that the cows can be automatically fed with small, easily digestable quantities of feed many times throughout the day. The system requires automatic cow identification, food storage bins and special feeder stations at strategic points in the cow housing areas. The operator has to key in the total quantity of feed that each cow should receive. This is done from the farm office. The total ration will then be automatically dispensed in small amounts over a twenty-four hour period.

Automatic cluster remover—With a correctly installed and maintained cluster removal system, the risk of overmilking is much reduced. If the teat cups are left on when milk flow has ceased then the delicate tissues of the udder can be damaged. The end-of-milking indicator is a major part of any ACR system and may be operated by a float, electrical contacts measuring resistance, or the ceasing of milk flow through the milk meter or recorder jar.

To compliment the ACR, automatic milk transfer can be used to open a valve in the transfer pipeline at the bottom of the recorder jar to allow milk removal. The valve closes when the jar is empty with little loss of vacuum. A manual override is normally provided to hold the milk for diversion or recording and sampling purposes.

Farm management systems run on personal computers (PC)—Although a certain amount of information can be gathered automatically (i.e. milk yield), the operator must be prepared to sit and input herd details into the system via the keyboard to make the management programme perform to its best potential.

A good system should provide at least:

—herd history (cow name, number, parents, number of lactations, days in present lactation)
—milking performance (amount of milk in present lactation, milk analysis)
—feeding details (calculation of daily ration of roughages and concentrates based on measured energy and dry matter values etc. and the actual and projected milking performance, operation of in- or out-of-parlour feeders with pre-programmed quantities of food)
—health details (veterinary visits, mastitis, pregnancy diagnosis)
—communications (warnings for operators when milking—should be two way contact i.e. parlour to PC and PC to parlour)
—exceptions (highlight cows which have dropped below projected performance levels, attention drawn to these animals on VDU screen or by hard copy from printer).

If the PC is not dedicated to running the dairy herd, then it could also be used for arable records, wages etc.

REMEMBER, total farm management systems are expensive to install and maintain; however there is nothing to stop a farmer purchasing small sections of the system i.e. ACR or semi-automatic feeders, which are affordable and suitable for the size of the dairy herd in question.

Milking by Robots

The use of the robot to milk cows is a very exciting and realistic proposition. There are already several robotic milking projects taking place on selected farms in Europe and this futuristic way of milking will soon be available for purchase or lease. The cow can be milked several times a day in a milking stall similar to those used for tandem milking parlours. After being automatically identified, the cow is fed and her udder prepared. The milking unit then senses the position of the teats and fits the teat cups. Milking then continues as normal, until milk flow ceases after which the cluster-remover takes the cluster from the udder. The teats are then given a post-milking dipping and the cow is released from the stall. The milk is filtered and goes to be chilled in bulk storage. As the cow is milked the opportunity arises to measure the conductivity and temperature of the milk for mastitis and heat detection. Milk composition can also be analysed. If further treatment is required the cow can be diverted to a separate holding pen. The cluster is automatically washed thoroughly and forced-air dried between milkings.

MILK TRANSPORT, COOLING AND STORAGE

Milk from the recorder jar, milk meter or milking pipeline is collected in the receiver vessel, which should be installed in the milking parlour pit. This eliminates the need for any milk to travel up vertical sections of pipework which is demanding of the vacuum supply. Careful consideration must be given to the diameter of the pipe to ensue that it is adequate to carry the quantity of milk required. The same pipework also carries the cleaning solution and incorrect sizing of the pipe will affect the performance of the cleaning cycle.

Since the air pressure inside the receiver is at milking vacuum it is necessary to use a special pump to transfer the milk to the

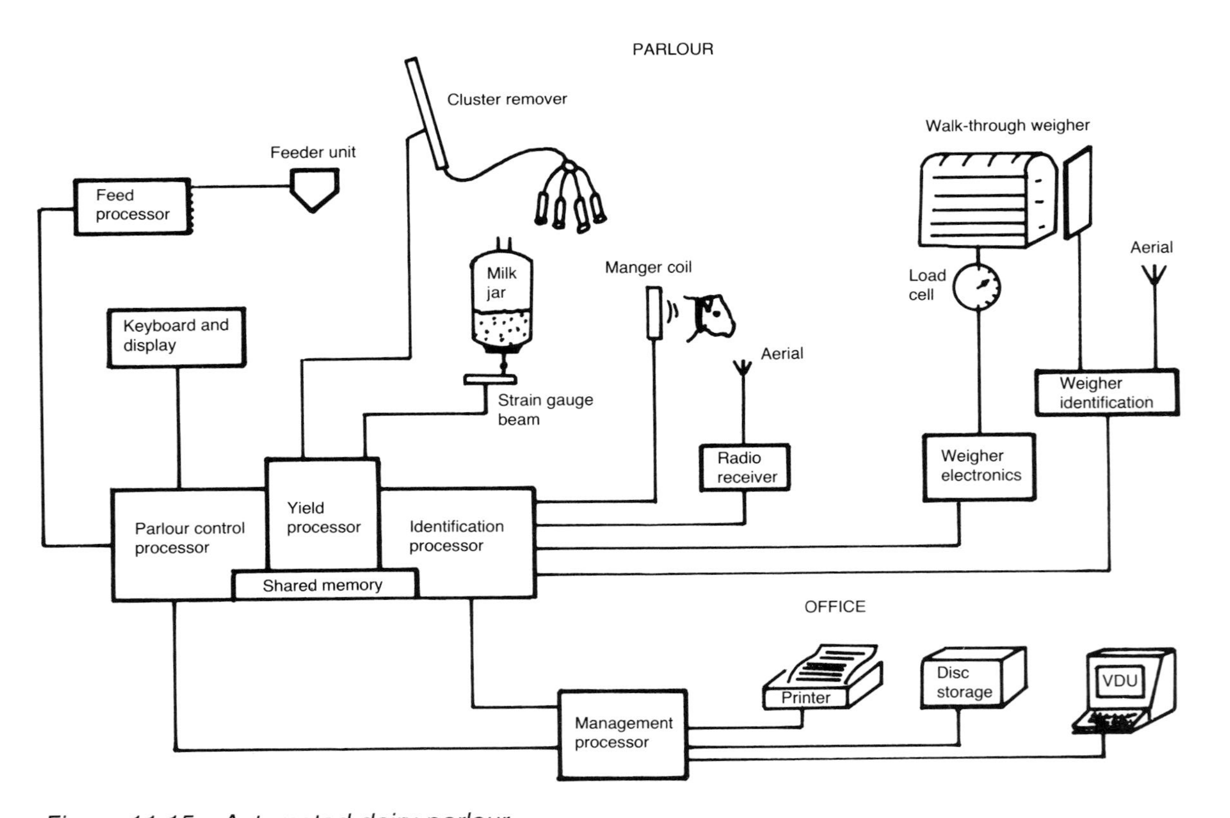

Figure 11.15 Automated dairy parlour
The British Society for Research in Agricultural Engineering

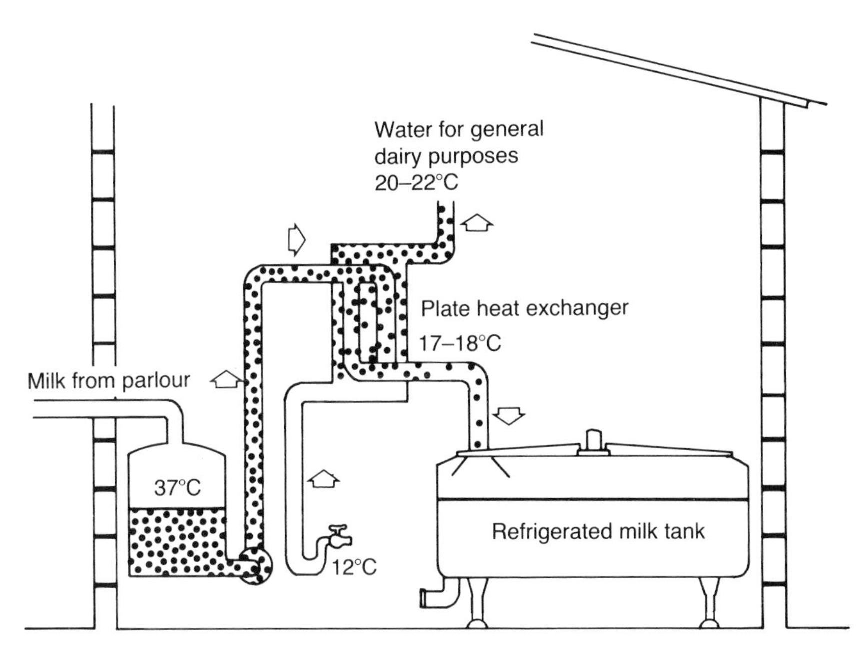

Figure 11.16 Plate heat exchanger
ADAS

storage vessel. This may be an electrically driven rotary pump or a pneumatically operated diaphragm pump. Once the milk leaves the pump it is at atmospheric air pressure.

After filtration the milk reaches the bulk milk storage tank, where it is cooled by water from an ice bank to a maximum temperature of 4.4°C. A small microprocessor can be used to control the tank's functions including making the best use of off-peak electricity. The heat which is lost when milk is cooled can be used to heat the water for udder cleaning, calf feeding or for cleaning the milking plant.

Pre-coolers use a series of heat exchanger plates or tubes to cool the milk to within 2–3°C of the incoming farm water temperature which is usually 10–14°C.

As the milk temperature is reduced, then the water temperature increases by 8–10°C. About 2½ times the volume of water to milk is required. If the water is bought from the mains supply, then it must be re-used and not run to waste, otherwise no savings will be made. In the winter the water can be stored and used for cattle drinking.

Heat recovery units use the waste heat taken from the milk to heat the water in an insulated tank surrounded by part of the condenser coil (Figure 11.17).

The temperatures achieved can be as high as 60°C. Further heating by an immersion heater is required to boost the water up to circulation temperature (85°C).

An alternative to an ice bank tank (indirect expansion) is the direct expansion (DX) type, where the evaporator tubes are bonded to the outside of the milk vessel and the heat of the milk is absorbed across the milk vessel wall by the refrigerant. This principle has allowed the manufacturer to make a more compact tank of circular or elliptical cross-section which provides extra storage capacity. The refrigerant unit only works when there is milk to be cooled and therefore a more powerful unit is required since with ice banks the refrigerant unit can work continuously at a low rate to produce the ice. This unfortunately increases the overall cost of the tank. The extra cost can be deferred by complementing the DX tank with a chilled water pre-cooler and the use of off-peak electricity.

All milk collection in the United Kingdom is now by bulk tanker so it is essential that the driver can measure the quantity

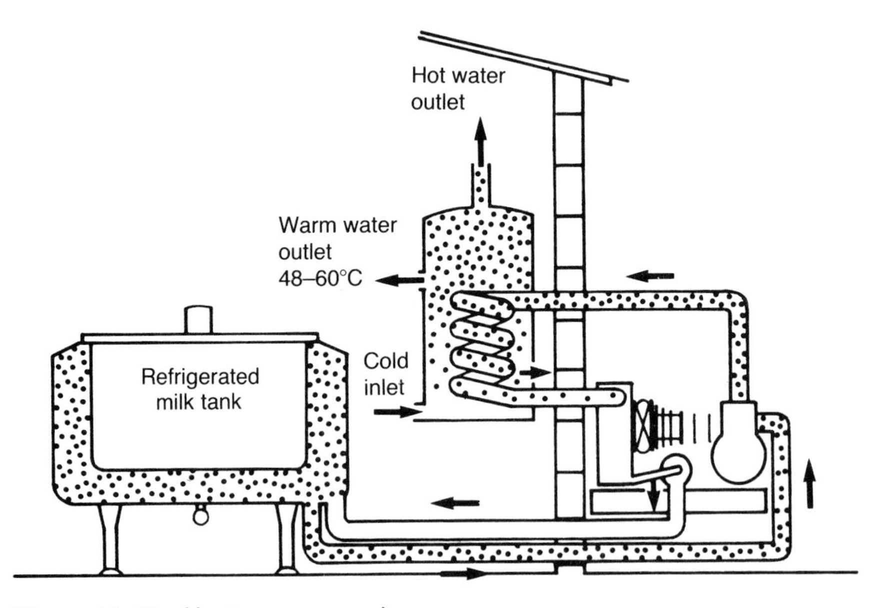

Figure 11.17 Heat recovery unit
ADAS

of milk accurately and can quickly take a representative sample for laboratory testing. These factors together with very stringent cooling requirements have lead to the types of bulk milk tanks which are currently available. There is an ongoing updating of all milk collection tankers for the fitting of milk meters for measuring the quantity of milk as it is pumped from the tank. Accompanying the meter is a small printer which prints a ticket to show the quantity of milk collected. A copy of this ticket is left for the farmer's records. The use of meters on the tanker provides manufacturers with greater flexibility of farm tank design since the tank no longer has to be a ridged calibrated vessel. The main types of bulk milk tank are illustrated in Figure 11.18.

Tanks are generally made from stainless steel and must comply with very strict standards regarding construction, refrigeration and agitation, making them costly items to produce.

TESTING OF MILKING MACHINES

The farmer expects complete reliability from his milking machine for up to 1,095 milkings per year. Furthermore the milking machine is the only farm machine to come into direct contact with animal tissue, and thus nothing short of the highest standards of maintenance and testing should be applied. Unfortunately, this is not always the case and the resulting gradual decline in performance may not be noticed by the herdsperson. Therefore, in order to keep a milking machine in the peak of condition, it is necessary to adopt a regular service schedule backed up by a thorough regular test (at least annually).

The method of testing is described in BS 5545 and since it requires specialist equipment, it cannot be carried out by the average farmer. Instead it can be undertaken by the Milk Marketing Board (Genus), ADAS or the Milking Machine Manufacturers, all of whom were involved in drawing up the standard test.

In principle, the test records the performance of most parts of the installation including the vacuum pump, regulator, pulsators, valves and pipe work. The results of the test are compared with standard figures and if faults are detected recommendations for their correction are given.

It would be quite wrong for the farmer to rely upon such testing to locate problems and he should adopt a programme of checks and preventative maintenance to ensure long and efficient service from the milking machine.

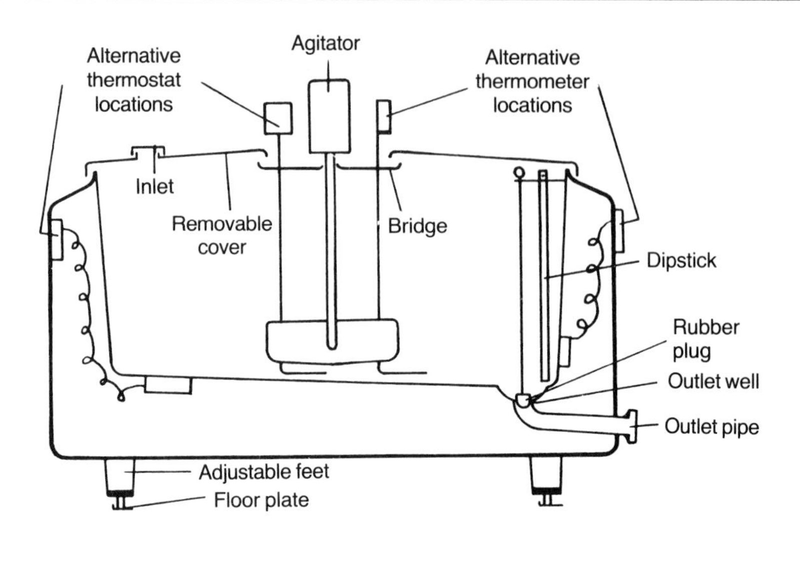

(a) (above) Typical tank showing features common to all

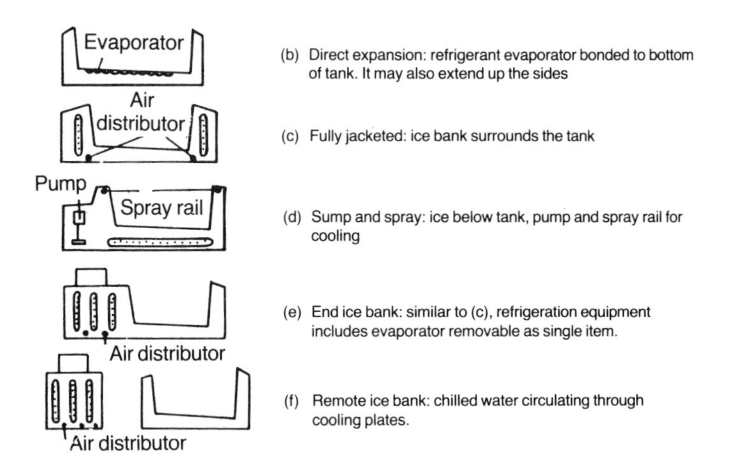

(b) Direct expansion: refrigerant evaporator bonded to bottom of tank. It may also extend up the sides

(c) Fully jacketed: ice bank surrounds the tank

(d) Sump and spray: ice below tank, pump and spray rail for cooling

(e) End ice bank: similar to (c), refrigeration equipment includes evaporator removable as single item.

(f) Remote ice bank: chilled water circulating through cooling plates.

Figure 11.18 Types of refrigerated bulk milk tanks
NIRD Technical Bulletin

This testing is called a 'static test' and is undertaken when there is no milk flowing through the plant. The testing agencies will also undertake a 'dynamic test' whilst milking is in progress. This test will highlight problems such as vacuum fluctuations brought about by poor operator technique or incorrect pipeline sizes.

Other tests which are available from the advisory services mentioned above include feeder calibration and an evaluation of the cleaning performance to BS 5226.

CLEANING OF MILKING EQUIPMENT

All surfaces which come into contact with milk must be regularly cleaned in order to maintain the hygienic quality of the milk and prevent contamination.

Bucket milking plants are cleaned by hand washing but recorder and pipeline machines may use either a circulation cleaning or acidified boiling water (ABW) method. It is recommended that all plants are subjected to the full cleaning routine after each milking, but some farmers use a simplified method after evening milking.

The Control of Substances Hazardous To Health Regulations (COSHH) MUST be observed when handling any dairy chemicals (or in fact anything else on the farm which may be dangerous).

Which cleaning routine is followed depends on the type of cleaning method used.

Circulation Cleaning

The plant is first rinsed with cold or tepid water in order to remove the bulk of milk residues and to make best use of the cleaning solution. For most plants 8–10 litres/unit is satisfactory.

The recommendations for circulation cleaning vary according to the chemical used but typical requirements are as follows;

1. Use 10–14 litres of water per unit heated to at least 85°C.
2. Draw hot water through the plant until the temperature of the discharged water reaches 50°C.
3. Connect discharge pipe to the wash trough to provide circulation and add the cleaning chemical to the wash water.
4. Circulate the water within the plant for 10 minutes and then discharge to waste.
5. Rinse with cold water (10 litres/unit), preferably containing 25 ml of hypochlorite per 40 litres of water.

Acidified Boiling Water Cleaning

This system uses near boiling water to clean and sterilise the plant and an acid to prevent hard water deposits.

(text continues on page 209)

Milking Machine Maintenance

Cheshire College of Agriculture

Daily Checks

1 Vacuum pump lubrication may need to be checked daily or weekly, depending on the type of lubricator used. The one shown introduces a controlled flow of oil to the pump interior via the bearings. High grade vacuum pump oil must be used for this job.

2 Vacuum gauge reading should be noted at each milking. The normal working vacuum level should be reached quickly and an audible hiss should come from the regulator.

3 The clusters should be checked for air leaks, particularly where pipes are joined. Any leakage at this point may cause severe reduction in milking efficiency, together with udder damage.

Claw piece air admission holes should be checked and unblocked if necessary—this is extremely important and should never be neglected.

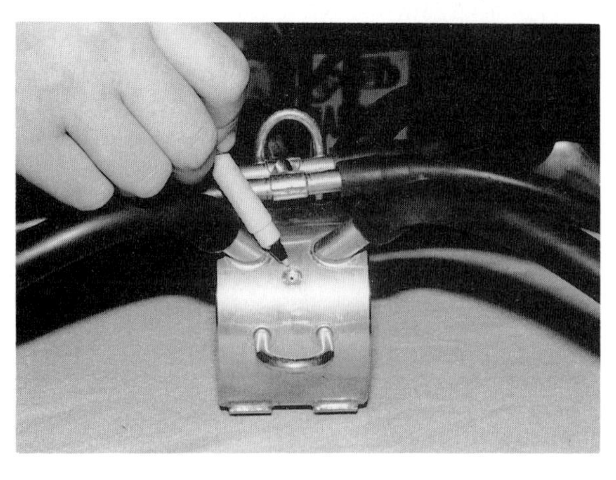

4 If a liner splits it will allow milk to pass through the pulsation system, into the air pipeline (pulsators) and into the interceptor. This requires prompt action to repair the fault and clean the contaminated parts. A visual inspection of all liners and tubing should be carried out weekly.

The milker should keep eyes and ears open for any unwanted air leaks, particularly where pipes are joined. These may be found when a full test is carried out, but constant vigilance is a much better method of detection.

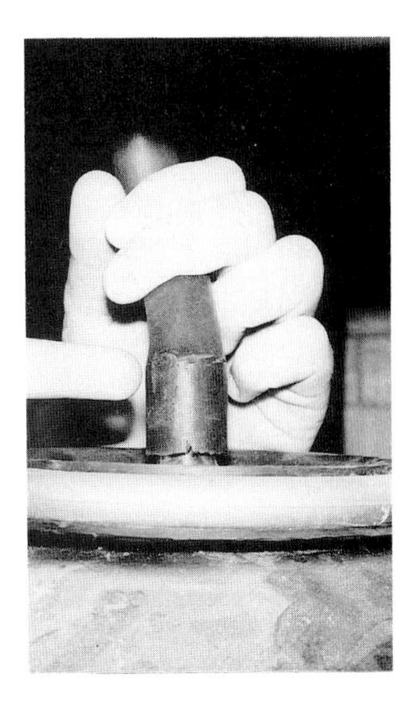

Weekly Checks

1 Vacuum pump drive belt tension should be checked by measuring the deflection of the belt when moderate force is applied mid-way between the two pulleys. This can be adjusted by moving the motor in its mounting slides. Insufficient belt tension causes belt slip, whereas excessive tension damages the belt and may overload pump and motor bearings.

2 Belt pulley alignment is very important and should be checked with a straight edge. If the flanges of both pulleys do not line up, adjustment is carried out by moving the pump and motor on the bedplate or by sliding the pulleys on their shafts.

3 Pulsation should be checked by inserting a thumb into the liner with all other teat cups plugged or cut off. Slow or indistinct liner action indicates air leaks, tube blockage or a faulty pulsator.

4 The rubber gasket fitted to the interceptor should be inspected for signs of damage or perishing.

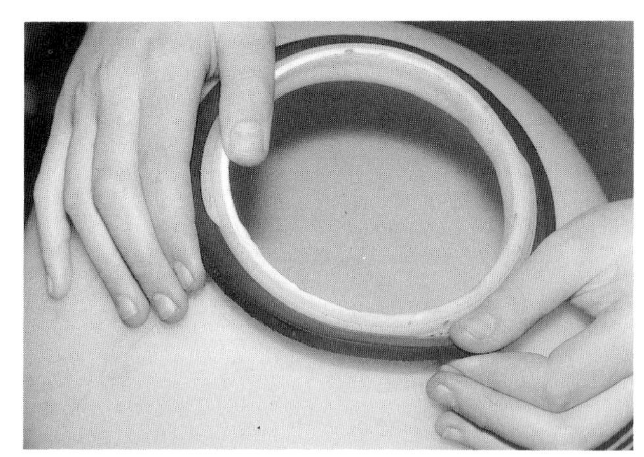

5 Examine teatcups for correct sealing of liner.

6 Check the operation of vacuum shut-off valves.

7 Clean condenser unit with soft brush or blast of air. Note: switch off mains before maintenance work commences, and switch on mains after maintenance.

8 Check operation of agitator on bulk milk tank.

9 Check milk temperature.

Monthly Checks

1 Clean regulator valves. Note that no petrol or spirits should be used on plastic components.

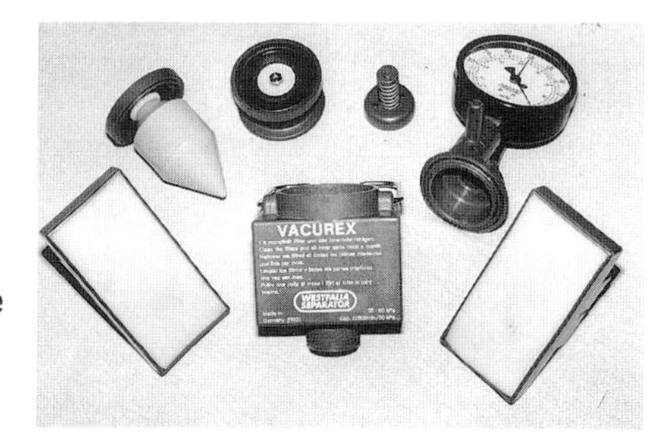

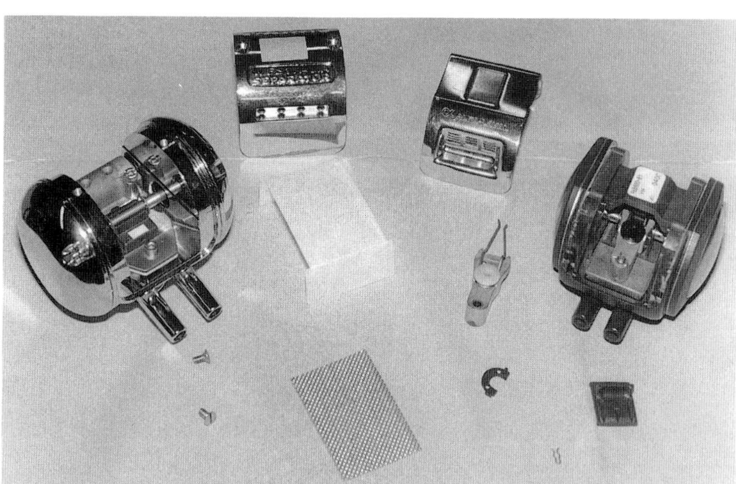

2 Pulsators should be dismantled, cleaned and inspected using the manufacturer's instruction. Particular attention should be paid to keeping airways clean and, if recommended, lubrication should be very sparing. The end plates of fluid-filled pulsators must *not* be removed (Alfa Laval Hydropulse).

3 Hot detergent disinfectant should be drawn through the air pipeline monthly. The rubber tube fitted to the end of the air pipeline should be lifted in and out of the cleaning solution to create a surging action for better cleaning. The quantity of liquid drawn through the pipeline should be approximately five litres less than the volume of the sanitary trap.

4 Check for milkstone in claw bowl and tube nipples.

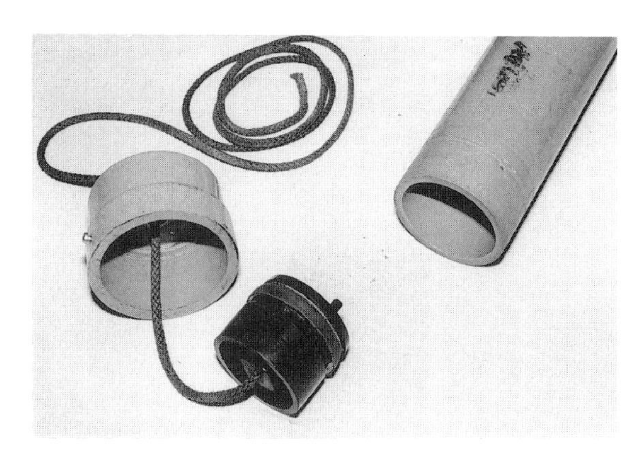

5 Lubricate take-off cylinder.

Six-Monthly or Periodic Checks

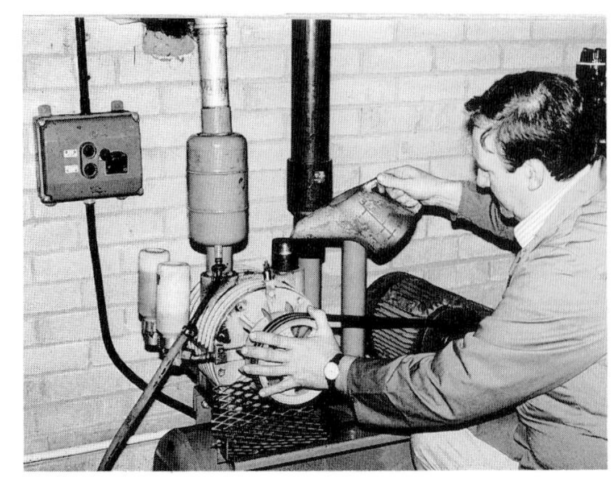

1 The vanes of the vacuum pump tend to stick in their slots after a long period of use. This can be corrected by flushing the pump. The method of flushing varies according to the type of pump and its exhaust system, so advice should be sought before attempting the job. The photograph shows paraffin being poured into the inlet of the pump—having first disconnected the electrical supply.

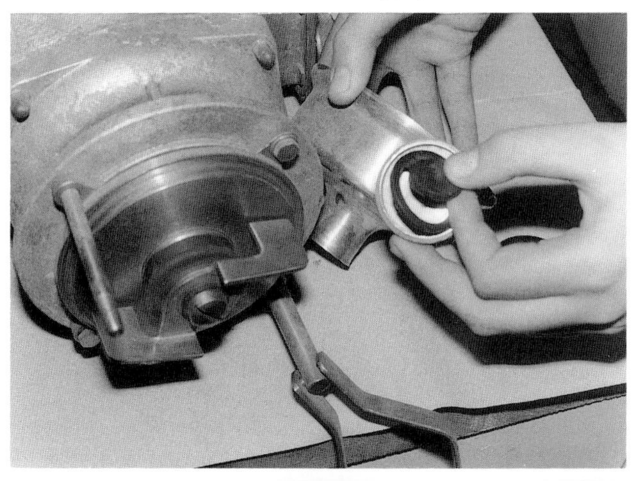

2 The milk pump may be fitted with a one-way valve to stop air entering through the pump when it is not operating. Air bubbles entering the receiving jar indicate this fault.

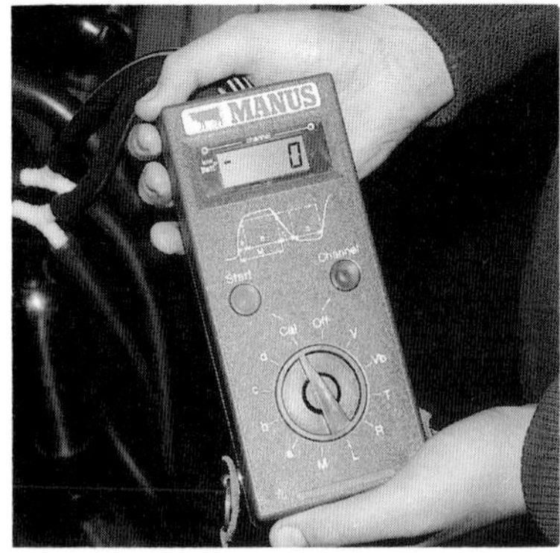

3 Accurate pulsator performance check.

A typical ABW routine is as follows;

1. Pre-rinsing is unnecessary.
2. Draw very hot water through the plant (14–18 litres per unit at 96°C), adding acid during first 2 minutes.
3. Discharge water to waste, ensuring that all parts of the plant are heated to at least 77°C for 2 minutes.

This system uses more water than circulation cleaner but it is quicker and the chemicals are cheaper. However, the cost of electricity to boost the water temperature to near boiling can be expensive. Great care must be taken when handling concentrated acids and protective clothing must be worn.

Descaling

Milk residues which are not removed in the normal cleaning process may build up to form a scale. This is best removed by soaking in 'milkstone remover' which may be based on phosphoric acid or other proprietory compounds. Care should be taken with aluminium and tinned surfaces, since these are attacked by milkstone remover.

MAINTENANCE OF THE MILKING MACHINE

All manufacturers provide maintenance instruction for their machines and these invariably involved daily, weekly, monthly and periodic tasks. The following sequence of photographs illustrates some of the principal items of maintenance, but it must be emphasised that all makes of machine differ in their requirements and there is no substitute for following the manufacturer's instructions.

CHAPTER 12

QUALITY-HUSBANDRY IN MILK PRODUCTION

Mention has been made concerning the differing qualities required in milk in order to obtain the best possible milk price. Quite apart from the extra payment attached to good quality, dairy farmers are obliged to produce milk of the legally required standard.

In 1985, the EC Health and Hygiene Directive was introduced to harmonise standards within the Community in order to facilitate trade and to maintain existing domestic standards.

The United Kingdom chose to market non-standardised whole milk. This is milk with a natural fat content of not less than 3.0 per cent. This legislation points out one of the first areas where dairy farmers can encounter problems since falling below the standard will prevent the milk from being sold at all. Milk which has low butterfat causes problems even if it is legal, since it receives only a very low price. Good butterfat contents are therefore essential.

CAUSES OF LOW BUTTERFAT

In the event of butterfat problems, the following possibilities exist:

Breeding. The butterfat percentage of milk varies as much between individuals of the same breed as it does between breeds. The production of butterfat is an inherent characteristic, so that low butterfat may be due to the breeding of cows.

The average yield, butterfat and protein percentage in the milk of the different breeds of recorded dairy cattle in this country, as published by National Milk Records, is given in Table 12.1.

The lowest of the recorded breed averages is approximately 3.8–3.9 per cent, which is well above the legal minimum.

Stage of lactation. During lactation, butterfat percentage falls as yield rises and vice versa. Consequently, the bulked milk of a herd is more likely to fall below standard if there are, at any one time, a majority of the cows at the peak of their yields.

Age of cow. A high proportion of old cows in the herd will tend to reduce the butterfat in the bulk sample since butterfat percentage generally declines with age. Heifers test higher than cows.

Feeding. It is generally accepted that changes in feeding, on an adequate and well-balanced diet, have only a minor and temporary effect on butterfat percentage. Certain foods like palm kernel cake or coconut cake tend to raise the percentage slightly, whereas other foods—for example, cod liver oil—depress butterfat.

A cow calving in good condition will give milk of higher butterfat percentage than if she calves very lean. On an inadequate diet, yield and butterfat tend to fall below normal.

On a highly laxative diet the butterfat percentage tends to fall. This often occurs in the spring when the cows go to grass. It can be remedied by:

- restricting the intake of grass by controlled grazing, or by turning the cows on to a more mature type of pasture;
- feeding some dry roughage food such as hay or straw, say 1.8–2.7 kg daily, before the cows go to grass in the morning. This method has been used for many years with high-yielding cows.

Table 12.1 Yield, Butterfat and Protein Breed Averages, England and Wales

	Milk Yield		*Butterfat*		*Protein*	
Recording Year	*1986–7*	*1987–8*	*1986–7*	*1987–8*	*1986–7*	*1987–8*
	kilograms		*per cent by weight*			
Ayrshire	5,173	5,184	3.97	3.96	3.33	3.32
British Friesian	5,740	5,751	3.91	3.89	3.22	3.22
British Holstein	6,431	6,443	3.87	3.86	3.16	3.16
Dairy Shorthorn	5,142	5,006	3.72	3.71	3.28	3.27
Guernsey	4,205	4,224	4.70	4.68	3.54	3.56
Jersey	3,972	3,978	5.35	5.34	3.81	3.82
South Devon	3,072	3,522	4.09	4.01	3.59	3.61

Low butterfat may be due to a low intake of fibre; this would be most likely to happen on young spring grass.

Frequency of milking Theoretically it is advisable to milk cows at a twelve-hour interval; in practice this is a system which is seldom adopted. With the normal period of ten to fourteen hours, the recorder is likely to find a higher butterfat level at morning milking than at the evening milking. Differences of up to 1.0 per cent have been recorded. With three–times–daily milking, milk yield is increased by up to 15 per cent. Butterfat percentage is slightly lower but an overall greater weight of fat is likely. The new payment schemes are concerned with total weight of fat and protein.

Health and welfare of the cow Cows that are off-feed or sick show a greater drop in milk yield than in butterfat percentage. Lack of exercise will also reduce butterfat percentage, as would undue excitement at milking time.

REMEDIES FOR LOW BUTTERFAT

The dairy farmer who meets the problem of low butterfat percentages has several husbandry practices which he can follow to improve butterfat.

Cows should calve down in fit condition and must be adequately fed without any tendency to scour. There should be adequate provision for exercise. Cows should be well housed. The bulking of milk in the tank averages the effect of milking interval. The provision of roughage—fibre—to the cow when on spring grass is important.

Another way of stimulating butterfat is to feed sugar beet pulp nuts, which have a medium energy level but a high level of digestible fibre and therefore promote butterfat production.

With an autumn-calving herd, higher yields, coupled with a high level of feeding can cause low butterfats. This is largely due to traditional twice-daily feeding of concentrates in the parlour, which tends to make the pH of the rumen fluctuate. At a low pH the acidity of the rumen considerably reduces the production of acetic acid which is important for the production of butterfat. The solution is to move towards multi-feeds of concentrates on a feed fence, coupled with an increase of fibre in the diets. The complete feed system also tends to satisfy some of these requirements.

If these measures fail to meet the situation, the problem is then likely to be associated with breeding. The routine monthly testing of butterfat will identify those animals which provide a constant problem. The programme for culling in the herd should eventually correct the faults but care will have to be taken in the selection of future bulls. These should be chosen for their ability to pass on better levels of butterfat production.

LOW SOLIDS-NOT-FAT

The main causes of low solids-not-fat (snf) are related to the management of the herd, rather than to the breeding. Although there is a basic difference between the breed types, e.g. Ayrshires tend to have a better figure for protein and butterfat than Friesians, feeding and management seem to largely dictate the actual levels.

Underfeeding, either in quantity or quality, for any long period, causes a significant decline in protein in many herds. This is most likely to occur in the late winter months when fodder stocks may be running low; it will also occur if the quality of the fodder is of poor quality as a result of bad weather at harvest time.

There is a similar danger in periods of drought; the cows may be underfed at pasture. Supplementary feeding should therefore be introduced at the right time.

A study of protein analyses indicates that there is a definite seasonal pattern. Protein levels are usually quite low in the late winter but rise to a peak in May and June.

The level is also related to the stage of lactation. In early lactation when milk yields are high, both protein and butterfat tend to be low, but levels rise as the lactation progresses—the protein level certainly improving as the cows move on to spring grass.

The variations in level are magnified when a herd calves on a 'block' basis; on the other hand if the herd calves over a long period of time the protein level tends to be stabilised.

It cannot be too strongly emphasised that adequate feeding of dairy cows is essential if the quality of milk is to be maintained.

A common mistake made by too many milk producers is to delay the supplementary feeding when cows are on autumn grazing. Early autumn- and winter-calvers need special attention at this time; if they lose condition too rapidly before the onset of winter, then trouble can be expected the following spring.

Mastitis is one reason for loss of milk production within a herd, and it also accounts for a low snf figure. As a sub-clinical infection it can cause a reduction in snf of the order of 0.2 per cent. This is mainly due to a drop in lactose and casein in the milk.

TAINTS IN MILK

Taints in milk arise either from foods eaten by the cow, or from odours absorbed by the milk from the atmosphere.

Taints from foods are most likely if feeding occurs just before milking and it is now a common practice to limit foods such as kale, swedes and silage immediately before the cows are milked.

Beet tops are particularly liable to taint milk if consumed in a semi-decayed state. They should be fed reasonably fresh, albeit after a wilting period of at least 48 hours.

Certain weeds also cause taints in milk—the chief one being garlic, a common weed of hedge bottoms.

Odours from disinfectants or fly sprays can cause taints which are absorbed from the atmosphere; it should always be a general practice to keep the dairy as fresh as possible.

VICES IN DAIRY COWS

Kicking The correct handling of the down-calving heifer will induce good milking habits. Initially, cows kick from pain or fear, and once the habit is formed it is very difficult to break.

The usual method of restraining a confirmed kicker is by passing a rope in figure-of-eight fashion around the hind legs above the hocks and pulling tight. This prevents either leg being used to kick at the milker. Another method is to pass a rope forward of the pelvic or hip bones around the abdomen and pull tight; this, however, is a less satisfactory method. Restraint in the form of a kicking bar can be used. This places a constraint upon the flank.

Always remember that the closer to the cow one stands while milking, the less harm she can do should she kick out.

The quiet but firm handling of cows is essential to good cowmanship. Where a cow kicks or is restless while being milked, consistent patience and gentleness are often more effective than forceful means of restraint.

Restraining methods should only be used after all other efforts have failed.

Suckling Occasionally individual cows exhibit a tendency to suckle. This habit is established in calfhood and is discouraged if calves are housed singly when pail fed. There are various devices on the market which consist of a plate which is fastened through the nose of the offender.

SLOW MILKERS AND LEAKING TEATS

Hard or 'slow' milkers are usually so because of unusually small teat orifices. The condition may be overcome by veterinary treatment using teat expanders, or by a delicate surgical operation which severs some of the sphincter muscles.

Such practices are always attended by the risk of mastitis infection.

The problem of leaking teats due to wide teat orifices can be temporarily met by sealing the teats immediately after milking by dipping them into collodion or by passing a rubber band round them.

Neither is a permanent cure and the best plan is to milk these animals more frequently. If a rubber band is used, care must be taken that it applies only sufficient pressure to stop the milk flow; it must not cause pain.

Pea In Teat

Occasionally an internal growth or 'pea' develops in the teat canal which makes milking slow and difficult. The obstruction may be removed by a veterinary operation.

Sore Teats

In cases of sore teats the pain caused while milking can be ameliorated by rubbing in udder ointment which softens the skin of the teat and promotes healing.

THE PERSONAL TOUCH

Before the subject of cow management is closed, I feel it is necessary to stress again how important the personal factor is.

The good herdsperson—whether male or female—should be

on terms of quiet confidence and affection with his cows. He should be able to approach them readily, even in the field, and at the same time be recognised as their friend. At no time should milking cows be roughly handled, hurriedly driven or shouted at.

A good herdsman is regular and meticulous in the work routine, as cows are essentially creatures of habit, and show resentment or nervous excitement when their daily routine is upset. If a cow is to be high-yielding then she must be contented.

The faculty of really observing cows, with a discerning and critical eye, is not a gift possessed by everyone, but for the first-class stockman it is essential. Early recognition of disease or symptoms of ill-health, even to the extent of just a loss of appetite, however slight, can mean a great deal if observed at once. A good stockman is a key worker on any dairy farm. Without such a person the best-planned schemes will never reach fulfilment.

'RECORDING HELPS'

Breeding

One of the less appreciated advantages of joining the official milk recording schemes is that record-keeping becomes a regular routine. The recording of calving dates, service dates and notes as to when cows are sick is provided for on the milk weighing sheets, and the Milk Record Book provides a summary of each cow's history during the milk recording year.

This is very valuable information. It is essential to the farmer in planning a breeding programme; it is essential to the stockman in day-to-day herd management, and it is especially useful to the veterinary surgeon when he is called in.

Where herds are not officially recorded some form of herd records should be maintained. The effort and time they involve will be more than fully repaid.

Many details will be found of significance as the records build up. For example, irregular heat periods, coming at less than 18 days or more than 24 days, must indicate hormone imbalance or mechanical failure of the reproductive organs. For this reason, it is desirable that all heat periods should be recorded whether the animal is served or not.

A recent introduction on many farms has been the Breeding

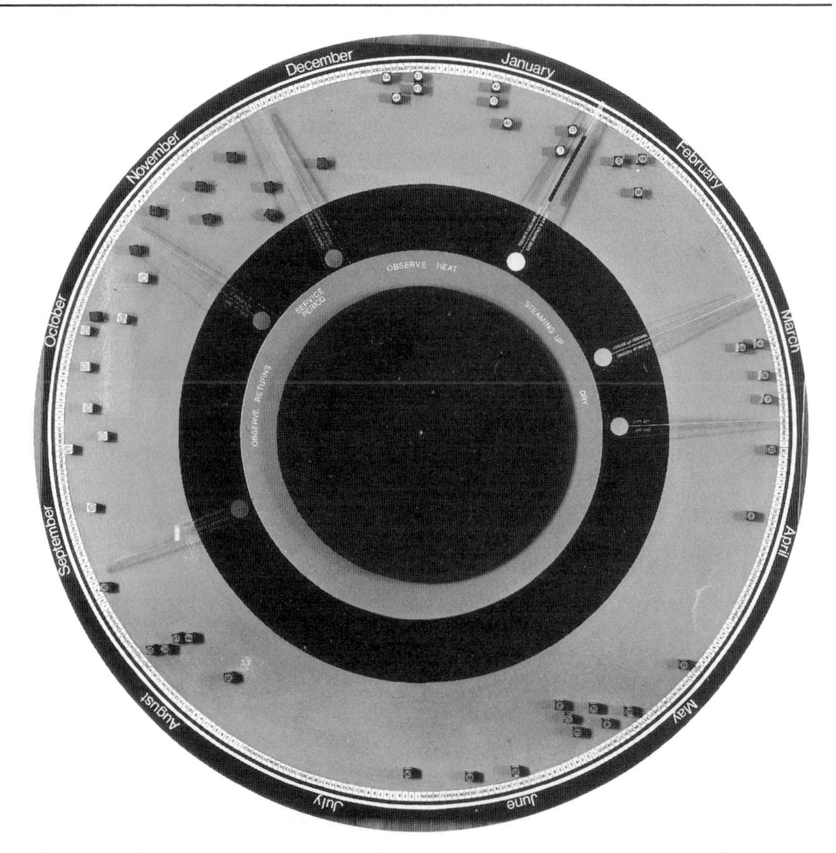

A breeding chart such as this can be fixed on the parlour wall or in the farm office and will provide at-a-glance details of how long each animal has been calved, when it is due for service, when it was confirmed in calf, when it should be dried off and steamed up, etc. Each cow is represented by a numbered magnetic cube.

Board. There are many types but they all perform the same task. The information provided by a breeding board will be:

- a record of all heats observed, giving the regular cycle of heat periods;
- the first service, and any recurrence of heat periods;
- second and further services;
- confirmation of pregnancy;
- the date of next calving;
- the dry period calculated with reference to next calving date.

With this information to hand the herdsman is keeping in touch with the breeding calendar of the herd on a day-to-day basis. Where cows are repeatedly failing to conceive to service, this is shown up clearly by the Breeding Board and veterinary help can be enlisted immediately. Ideally the veterinary surgeon should be used on a regular consultative basis; the herd health should be discussed regularly and in many cases problems identified before they become too serious.

Food Recording

One form of recording that is particularly valuable because of its relation to the cost of milk production is food recording. It is as important to know how much food the herd consumes as it is to know how much the individual cows yield.

For example, consider a 120 cow herd which has 90 cows in milk, producing 8,172 litres in a particular week, with the first 4.5 litres from each cow coming from bulk and the rest from concentrates at the rate of 0.45 kg/litre. Milk from bulk = 4.5 litres × 90 cows × 7 days = 2,835 litres milk.

- Yield from concentrates = 8,172 − 2,835 = 5,337 litres milk.
- Amount of concentrates needed to produce this = 5,337 × 0.45 = 2,401 kg.
- Concentrate for herd followers = 381 kg.
- Total concentrate usage for the week should be 2,401 + 381 = 2,782 kg.

The next step is to make sure that this is the amount of concentrate which was in fact fed. If there is a discrepancy, then it must be investigated.

Problem of Bulk Feeding

Silage is not easy to ration. For one thing its dry-matter content is highly variable. Nevertheless, a real attempt should be made to judge the weight fed, by weighing the occasional forkful and allowing for any variation in dry-matter percentage.

Roots like mangolds, swedes or beet tops are often fed at pasture when individual feeding of the cows is impossible. Nonetheless the weight in a cart- or trailer-load should be estimated from cubic capacity.

Kale, or other crops rationed by electric fencing, should be check-weighed before folding begins in order to determine the crop yield. Due allowance should be made for waste.

It is only if the actual quantities fed are known that a check can be made on the quality of home-grown foods. If the rations fed prove in practice to be inadequate, as would be shown by an excessively rapid decline in milk yield, or by loss of condition, then the quality of the home-grown foods has been over-estimated.

The cow is the best judge of their feeding value, but unless actual consumption is known the farmer is in no position to judge their adequacy. Home-grown foods exhibit a much wider range in feeding value than purchased concentrates.

Therefore, feeding of bulk feeds, employing high proportions of home-grown foods, needs intelligent stockmanship if results are to be satisfactory. Rule-of-thumb methods will not do.

The old concept of 3½ lb of concentrates per gallon (0.35 kg concentrates per litre of milk) has no equivalent when it comes to feeding silage and other bulk foods. The farmer has to make every effort to try and assess the value of his own material in terms of weight required for maintenance or per litre of milk.

Analyses can be of great help, but energy and protein details alone are of little guide to the feeding value of the grassland products such as silage, hay or dried grass, unless the fibre percentage—indicative of the stage of growth of the conserved material—is also known.

In the future, laboratory assessment of the feeding value of feeds will include a measure of digestibility which, as already pointed out, falls markedly from around 70–80 per cent with young conserved herbage and concentrates to 50 per cent or below for mature fodder such as cereal straw.

CHAPTER 13

BREEDING BETTER COWS

The continued genetic improvement of dairy cows involves a combination of skills and resources. The herd owner has the ultimate choice of parental mating combinations. However, the Milk Marketing Boards and other Artificial Insemination organisations undertake the major responsibility for progeny testing, milk recording, data processing and administration of the AI services. The Breed Societies' role in improvement is concentrated on correct identification of ancestry, conformation yardsticks, maintaining breeder interest, breed development and promotion work. The Ministry of Agriculture, in addition to involvement in research and development of artificial insemination work, also operates to maintain livestock quality control, particularly with regard to approving bulls for semen collection and regulations governing importation.

The objective of all concerned with the task of improving the dairy cow, irrespective of breed, is to produce a more efficient dairy animal. This obviously assumes that greater efficiency is linked to greater economic returns. This objective incorporates an inherent ability to produce large quantities of milk of high quality, by the efficient conversion of available feedstuffs, on a lactation and lifetime basis with the minimum of disease incidence. Although with the imposition of quotas, interest in high yields may lessen generally, there will still be many farmers who will reorganise their dairy herds to make use of cows with high yields at economic levels of feed input.

In the last few years a number of new concepts have been introduced to provide better 'tools' for the breeder to achieve his aim. The purpose of this chapter is to introduce and explain some of the technical background to animal breeding and clearly define its objectives. Strategies for implementing genetic improvement on a national and individual herd basis are also described.

INTRODUCING GENETICS— THE SCIENCE OF ANIMAL BREEDING

Each body cell of an animal contains a nucleus in which thread-like structures called *chromosomes* are located. The cow (or bull) possesses thirty pairs of chromosomes in each cell; every pair containing one chromosome inherited from each parent.

The genetic make-up of an animal is controlled by numerous *genes* which are located as bands along the chromosomes at specific *loci* so that the genes are in corresponding pairs. The genes for one trait (such as hair colour) may be identical in which case they are called *homozygous* and, for example, both transmit the character of red hair. If they are not identical (one transmitting red hair, the other black hair) the combination is termed *heterozygous.*

During the formation of a sperm or an egg the number of chromosomes is halved. Each parent therefore contributes one of each pair of genes towards the genetic make-up (or genotype) of the offspring.

The physical form (or phenotype) of the offspring will be determined by the combination of the genes from each parent, taking into account whether these are *dominant* or *recessive*. As the names suggest, dominant genes control a characteristic in preference to recessive ones.

Figure 13.1 shows the effect on the next two generations of the mating of two animals which are homozygous for contrasting aspects of one characteristic. The cow has two recessive genes producing horns; the bull has two dominant genes producing a polled animal. They produce a first generation in which the four calves are heterozygous. Although their genes are mixed the cows are all polled because they have each inherited a dominant gene for that characteristic.

When two of these polled but heterozygous animals mate, the resulting second generation (taking a broad sample of a large number of animals) will typically consist of three animals which are polled and one which is horned. The horned animal is the result of a homozygous pairing of the recessive genes.

A set of simple inheritance rules based on this kind of information was first proposed in the nineteenth century by Gregor Mendel, an Austrian monk. In dairy cow breeding the genetics is far more complex than this simple example and it must be borne in mind that an animal's genetic potential is determined by

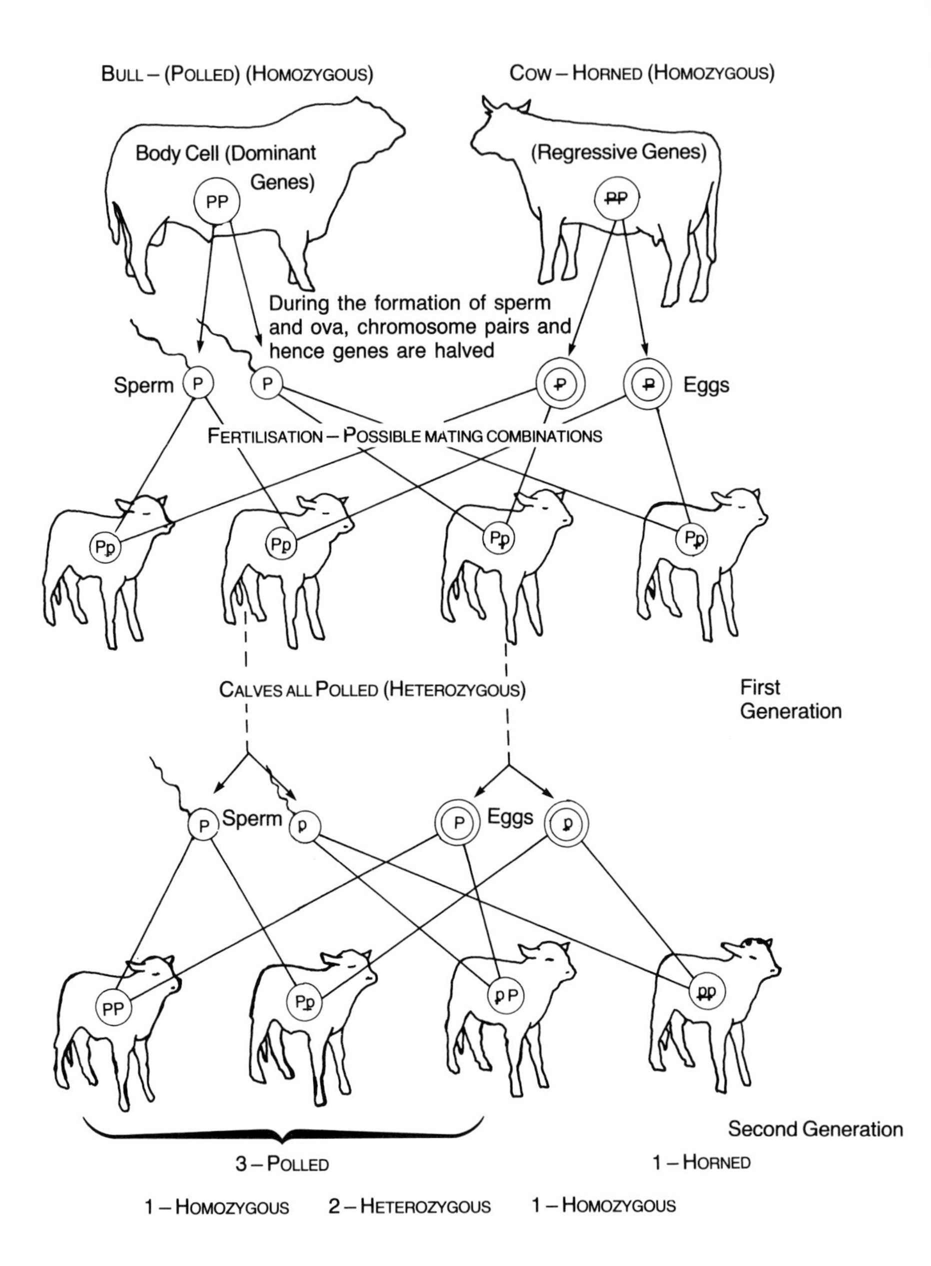

Figure 13.1 The effect of dominant and recessive genes

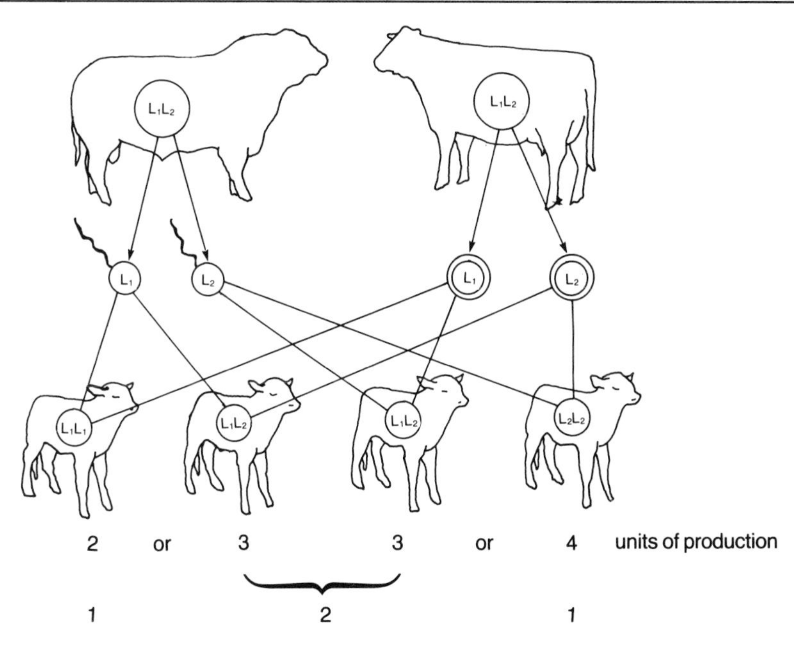

Plotted as a histogram this can be represented as shown in Figure 13.3

Figure 13.2 The inheritance of traits of economic importance

many pairs of genes all contributing a small proportion towards improvement.

Attempting to bridge this gap between the genetics of Mendel and population genetics is not easy but is very important for an improved understanding of breeding better cows.

This is best illustrated by considering a pair of genes carrying the potential for one or two units of production and acting additively, that is, they combine to manifest a production level rather than dominate one another, as described in Figure 13.1. This approach is shown in Figure 13.2. Different level of production potential are symbolised by L_1 and L_2. Each parent carries a gene which will determine a particular production level. During the fertilisation of eggs by sperms various combinations of productive potential emerge. The actual distribution of genotypes resulting from only a single pair of genes combining their productive potential is shown in Figure 13.3. This is further developed in 13.4 to exemplify a situation in which a production character is controlled by four pairs of genes.

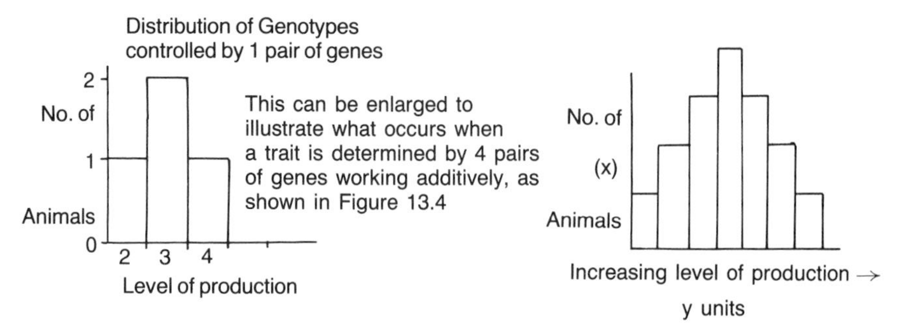

Figure 13.3 *Figure 13.4*

From genes to production curves

Most traits of economic importance in dairy cow improvement are under the control of many pairs of genes which have small effects which work additively to determine the genetic potential. Unfortunately, unlike in the simple example of animals being horned or polled we cannot usually see the effect of the genes and therefore cannot identify the genes, good or bad, that a particular animal carries.

A situation actually encountered in dairy breeding is shown in Figure 13.5 where the effect of many genes influences the yield of butterfat for a recorded population of dairy cows.

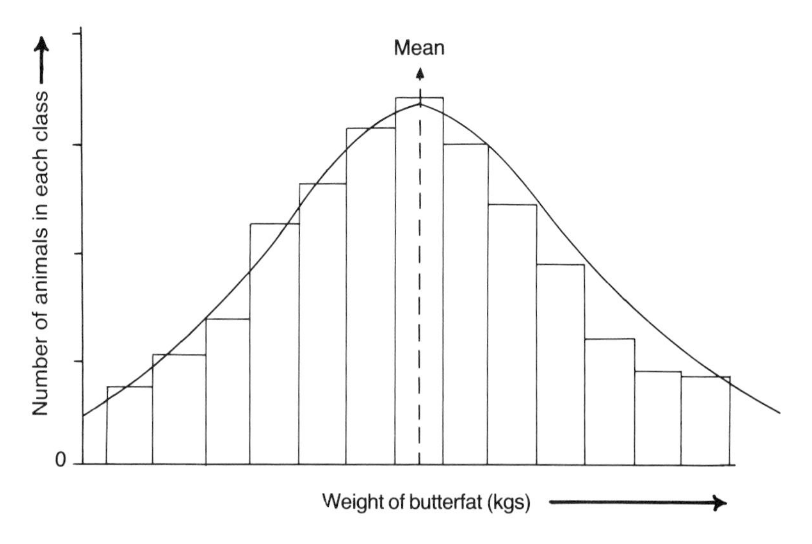

Figure 13.5 The distribution of weight of butterfat/lactation for a population of recorded dairy cows

Figure 13.5 shows how a shape eventually emerges to represent the productive level recorded in a particular population. Recording in this instance is the only way of identifying the effect of various genes. The geneticist would tend to represent a particular situation as a population of genes, and attempt to identify and select the best genes to produce the next generation and to cull out the poorest types.

The many pairs of genes which act additively to determine production levels have numerous effects which ultimately influence this level. For example, the genes for high production ability ultimately determine digestive capacity, heart and lung function, metabolic activity, udder conformation, etc. It is apparent therefore that selection for increased milk production also determines the conformation and constitution required to achieve this.

In order that we may identify animals with superior genetic potential within a population and select these as parents, it is necessary to understand some of the properties of this curve to form a theoretical background. This is illustrated in Figure 13.6.

This shows that the phenotypes represented in a population can be illustrated simply as a distribution graph. This population will have a mean value and a spread of phenotypes on each side of the mean. The objective of breeding is, very simply, to ensure that the mean performance of the offspring is better in the next generation. It is important therefore to identify the correct parents to produce the next generation. The shape of the population curve can vary for different traits but in all cases a definite proportion of the population will be within certain limits of the mean as shown in Figure 13.6. The 68% to 95% spread in Figure 13.6 define the area in the population curve which is the

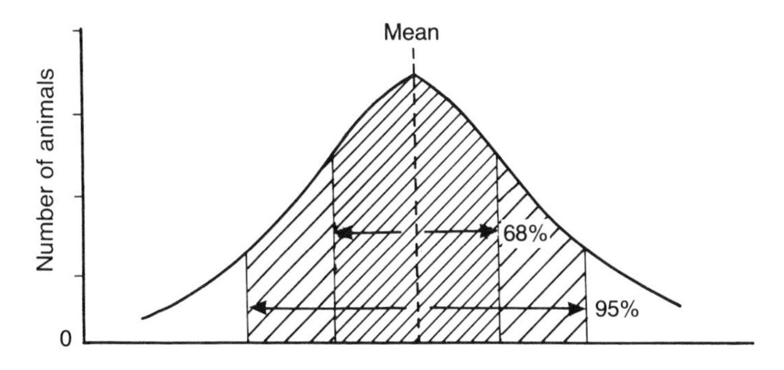

Figure 13.6 The shape of the population curve representing a phenotypic characteristic, e.g. milk fat production

basis for selection of different traits. The objective of the breeder therefore should be to select parents of the next generation which are considerably higher in genetic potential compared to the population average. The higher the performance of the individual relative to the mean of the population the more intensive is the selection. For example the MMB of England and Wales select some 300 cows as potential bull mothers from a total of 1 million recorded cows.

Although the example quoted in Figure 13.6 is theoretical it does form the basis of practical animal breeding. For example, if the bulls selected for breeding are not significantly better than the average population, then there will be no genetic improvement. In addition, if particular traits have low heritability (see below) then genetic improvement will be slow. Out of the 300 cows the MMB select for planned mating, some 120 young bulls will be tested with the aim of reintroducing 1 in 6 to the proven stud. Therefore of the original 1,000,000 cows only 20 will have produced sons which are introduced to the proven stud; this is a very high selection intensity. The national breeding programmes adopted thus have very definite principles of population genetics as their basis.

IDENTIFYING THE COMPONENTS OF VARIATION

Phenotype is the observed physical form or measurable performance of the animal, e.g. milk, eggs, liveweight gain. The range of values obtained from recording a population is referred to as the phenotypic variation (see Figure 13.7). The phenotype of an animal is determined by its genetic make-up (genotype) and the living conditions or environment to which it has been exposed.

$$\text{Phenotypic variation} = \text{Genetic variation} + \text{Environmental variation}$$

Thus, before we can commence any genetic improvement strategy nationally, we have to devise a recording system to determine the phenotypic variations. From such records it is possible to estimate what proportion of the total phenotypic variation is due to genetics and what proportion is due to environment.

The proportion of total phenotypic or measurable variation that is due to genetics is referred to as the heritability in its broadest sense.

$$\text{Heritability} = \frac{\text{Genetic variation}}{\text{Phenotypic variation}}$$

The narrower the range of genetic variation, the lower the heritability. Importation of animals into the population serves to widen the genetic base, giving increased heritability and hence a greater rate of genetic improvement. Estimates of heritability vary from population to population and should not be used out of context. The estimates range from 0 to 1 or 0 to 100 per cent. In Figure 13.7 two ranges of genetic variation are shown in order to illustrate that within a population the scope for selection does vary as defined by heritability.

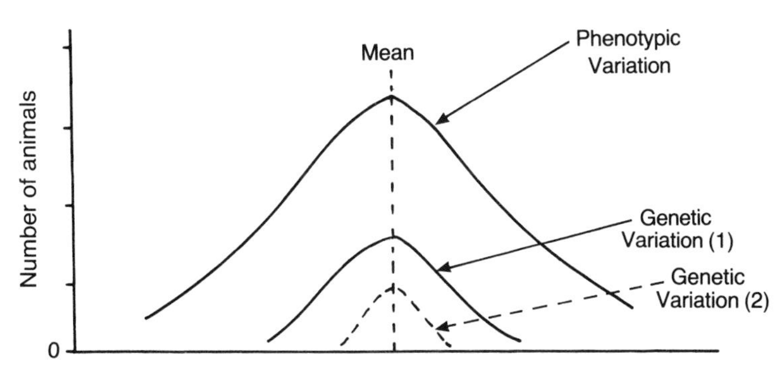

Figure 13.7 Proportion of the total variation due to genetics

SELECTION FOR IMPROVED PERFORMANCE

Selection refers to the process of identifying superior genotypes and using them as parents of the next generation. The master breeder is therefore concerned with the future and tries to produce animals one or two steps ahead of the majority of the population. The negative element of selection refers to identifying animals to be culled from the breeding population.

The principles which should dictate the choice of selection criteria have been thoroughly researched and may be summarised as follows:

1. Closely related to objective.
2. Capable of being measured easily, quickly, accurately and cheaply.
3. Capable of being measured early in life, if possible before sexual maturity, and in both sexes.

Selection Methods

The practical dairy breeder adopts many of the theoretical methods of selection which have been developed by the geneticist. The breeder chooses animals out of the herd as potential parents on the basis of their own *individual performance*. He aims to reduce the risk of selecting incorrectly by selecting an animal on more than one performance record and therefore considers the animal's *life time performance* as being a more reliable judgement of the ability for sustained high production (Figure 13.8).

The breeder can also derive a guide to an animal's breeding value by reference to the pedigree information. Information from families is more reliable if the relative importance given to each is correct; such combined information is referred to as a *selection index* and will be discussed later. Reference is made regularly to

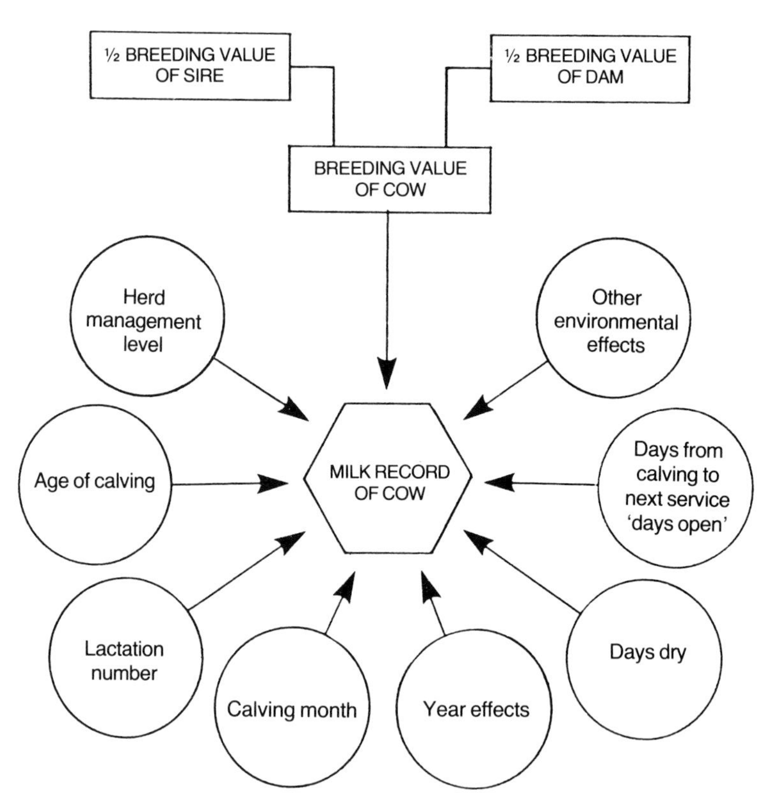

Figure 13.8 The components of the milk record of a cow
MAFF

cow families in pedigree breeding circles. The use of family selection takes many differing forms. Firstly, sire families, where a number of females are born of the same sire; this is usually termed as the progeny test. Secondly, cow families, where a number of half sisters out of the same cow, but with different sires, provide clarification of her breeding value. As the use of embryo transfer becomes more commercial, methods of cow progeny testing may become more sophisticated in the future. Great care should be taken not to attach too much value to the more orthodox use of cow families, whereby a whole number of animals in a herd are registered under the same name for several years. There are good reasons for exploring the ancestry for two or three generations, but only if this is allied to the objective methods of evaluation which exist. The basis for most selection decisions in the modern dairy herd is the progeny performance of individual bulls. This has become the cornerstone of national and herd breeding strategies.

Selection for More than One Trait

Most dairy breeders are pursuing an improvement in more than one particular trait over the same period. The three most common methods for selecting multiple traits are discussed below.

Firstly, *tandem selection*. In such instances the breeder concentrates improvement on one character until it reaches an acceptable level and thereafter concentrates on improving a second trait. This policy has to be pursued with care because:

- Improvement achieved with the first trait must at least be maintained when selecting for the second trait. Only animals which exhibit a good first trait should be considered for the second.
- The nature of the mechanism of inheritance may be such that improvement in the second trait leads to a deterioration of the first trait.

The second procedure is referred to as *independent culling levels* and is illustrated in Figure 13.9.

The breeder sets a threshold level for trait A, and notes any animals which are below this level; similarly the procedure is repeated for trait B. The animals located in the extreme left corner of the diagram, below the culling threshold for both traits will be culled. Figure 13.9 (ii) illustrates how the same procedures can be used to identify and thus select the superior animals.

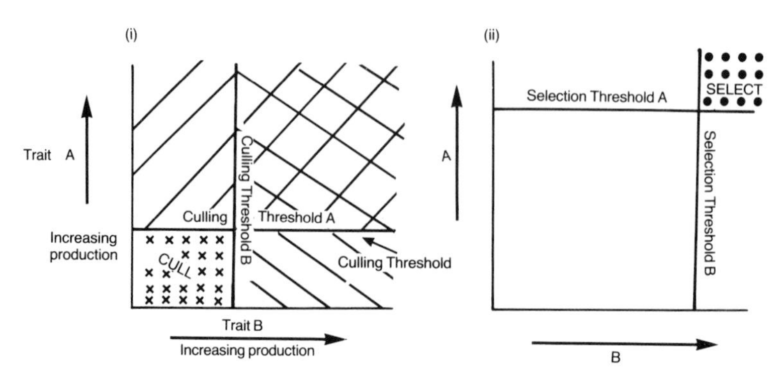

Figure 13.9 Selection and culling based on setting independent levels

The third approach is to combine all the information into the form of an *index*. This is illustrated by reference to a commonly used example.

Milk yield × % butter fat
= weight of fat (kgs)
Milk yield × % protein
= weight of protein (kgs)
} = Total weight of fat and protein

In the example, five different indicators of production level are combined into one selection objective. The breeder should therefore select for total weight of fat plus protein.

Inbreeding and Line Breeding

Inbreeding is defined as the mating of individuals who are more closely related than the average expected relationship of an animal in a herd. Line breeding is defined as mating a herd of cows to a succession of related sires, the line being chosen for a particularly good sire way back in the history of the breed.

Both these breeding methods were frequently used prior to the development of very accurate progeny testing methods. Their adoption in modern day breeding is more co-incidental than deliberate, in that the large scale use of proven sires result in a large number of relatives in a population. In addition, bulls with very high plus ratings for production traits will have a higher proportion of plus rated sons which means in fact that an element of line breeding and inbreeding will occur within the high producing strains.

DAIRY CATTLE BREEDING—SETTING THE SELECTION OBJECTIVES

In order to pursue a logical selection policy it is necessary initially to identify the traits which are to be improved and also their relative importance with the breeding policy. The greater the number of traits we attempt to improve, the slower will be the progress in each. Since genetic improvement is a long-term process, care must be taken to ensure that the objectives are correct and justifiable.

Broadly, our breeding aims are to produce cows of high production ability on a lactation and lifetime basis, which breed regularly, are healthy and have the temperament to withstand modern dairy herd management routines. Certainly under UK conditions the ability to produce efficiently from grass and conserved forage is very important.

Milk Yield and Milk Quality

Ideally we seek improvement in efficiency of food use for production of milk and milk solids. Measurement of efficiency on a national scale will be very costly. It is assumed at present that selection for higher production levels will lead to improved efficiency.

Most dairy producers make more money from selling milk than from surplus breeding stock. Therefore breeding for improved production has to be top of our priority of improvement objectives. The milk payment structure is being directed more towards milk solids production as shown on page 25 and therefore we should increase our breeding objectives in this direction.

The economic values reflect current market prices and therefore are liable to change over time. Weight of fat and weight of protein are eight and seven times more valuable, respectively, than the weight of lactose and thus should have the same relative priority in the breeding programme. Little information on lactose is available on individual cows or sire progeny groups at the time of writing, as it has only recently been included as part of our national recording or sire evaluation programme.

Heritability is a measure used by breeders to show the ability of a characteristic to respond effectively to selection. The scope for improvement of these traits is shown as the estimates of heritability (see Table 13.1).

Table 13.1 Heritability Estimates for a Range of Traits for the Friesian/ Holstein Breed

Trait	*Heritability*	*Trait*	*Heritability*
Production		*Conformation*	
Milk yield	0.31	Head, neck and shoulders	0.21
Fat yield	0.34	Body capacity	0.36
Protein yield	0.28	Top line and rump	0.31
Fat per cent	0.41	Legs	0.29
Protein per cent	0.44	Feet	0.28
		Fore udder	0.28
		Rear udder	0.32
		Teat shape	0.37
		Teat position	0.40

Source: MMB.

Note:

1. Heritability estimates are only relevant when discussing the population from which they derive.
2. Heritability estimates will vary for different breeds.

Table 13.2 Predicted Responses to Selection in Dairy Cattle

	Predicted response in daughters				
Criteria for bull selection for ICC	*Milk yield (kg)*	*Fat yield (kg)*	*Protein yield (kg)*	*Fat %*	*Protein %*
Milk yield	+++++	++++	++++	−−	−−−
Fat yield	++++	+++++	++++	+	−
Protein yield	++++	++++	+++++	−	−
Fat+protein yield	++++	+++++	+++++	0	−
Fat %	−−	++	−	+++++	+++
Protein %	−−−	−	−	+++	+++++

Source: MMB.

Table 13.2 demonstrates the predicted response to selecting bulls on the basis of any one of six different criteria. The figures represent the changes which would be expected in a herd of 'average genetic merit' if bulls were selected on Improved Contemporary Comparison (ICC) for one factor alone.

It is clear therefore that selection for fat plus protein yield is economically the preferred option. Greater emphasis can be given to milk yield, per cent butterfat and per cent protein in achieving the primary selection objectives of improving fat plus protein yield; such tactics will be dictated by the genetic status within a herd for the various straits.

Interpretation of sire progeny test information may assist dairy farmers in being more objective in choosing sires for use in their herds. Generally speaking, good levels of milk yield, fat and protein are all desired. If one trait, e.g. fat, is deficient then the primary objective is to improve the fat level whilst maintaining the yield and protein levels.

Table 13.3 Economic Values of Sire Progeny Evaluation

			Economic value		
	Fat weight (kgs)	*Protein weight (kgs)*	*Fat (£)*	*Protein (£)*	*Total (£)*
Bull A	+20.9	+12.1	45.35	26.74	72.09
Bull B	+11.6	+ 9.3	25.17	20.55	46.25
Bull C	+ 2.5	+ 1.8	6.42	3.98	9.40

Note:

Bull A: Example of one of the top bulls available in UK.
Bull B: Represents bulls available in top 1 per cent of Friesian/Holstein breed.
Bull C: Represents average merit of all sires of NMR recorded heifers in 1989. There will be an extra feed cost against this extra production.

1 kg fat = £2.17, 1 kg protein = £2.21.

The table shows the improvement associated with using Bulls A, B, and C. The physical improvement in kgs of fat and protein is shown as is the considerable financial advantage to be gained from the improvements.

Selection for Longer Productive Herd Life

Nationally, the average length of the herd life of dairy cows is disappointing. An increase in herd life would increase profit since older cows produce higher yields. In addition the number of replacements would be reduced releasing the land for other productive purposes, increasing the scope for selection and providing more opportunity for crossing with beef bulls.

Table 13.4 illustrates the difficulties in defining longevity. Criteria for measurement could be age, number of lactations completed, or total production of milk, fat or protein. The dairy farmer is concerned primarily with the productive utility of a cow, over a period of time under practical farm conditions and hence sustained production ability is very important. However, some dairy cows produce more milk over fewer lactations than others.

It is suggested that the total lifetime production of fat and protein should be our selection criteria. Any farmer culling cows that have given over 4 tonnes (4,000 kgs) of fat and protein on average may feel well pleased with his cows.

Table 13.4 Some Examples of Lifetime Performance of Dairy Cows

Length of herd life	*Lifetime production totals*		
Lactation no.	*Milk yield (kg)*	*Weight of fat (kg)*	*Weight of protein (kg)*
12	59,386	2,098	1,843
9	57,486	2,137	1,899
9	47,168	1,659	1,518
8	59,088	2,404	1,973
8	66,773	2,068	1,865
7	54,836	2,138	1,869
5	40,858	1,675	1,378

Lifetime profitability is a function of production per lactation, length of productive life and input and output prices.

The modern dairy farmer makes exacting demands on his cows and is quite prepared to expect maximum output for only a short life. Longevity must be earned on the farm thus, high-yielding young cows are likely to be allowed to stay in the herd longer than low-yielding cows. Therefore it is important to identify the potential high lifetime producer at an early stage so that her offspring can be retained. Large scale analysis shows that high heifer yield and high sire ICC rating are as accurate in predicting potential herd life as all conformation traits combined. Production and conformation differences combined however, only explain some 30 per cent of the variation in lifetime production.

Selection for Improved Functional Type

The importance of conformation in a breeding improvement programme is an emotive subject. Justification for inclusion of some conformation traits in breeding programmes revolves around their value in predicting productive ability.

There is no doubt that milk records are the most accurate assessment of productive ability. However, it it important to include some conformation traits in a breeding programme irrespective of whether they are related to productive ability or not. Traits associated with mammary structure, legs and feet have been shown to definitely be important in determining longer productive life in the herd.

Body Size and Efficiency of Feed Utilisation

It is generally assumed that bigger cows produce more milk over the fixed feed requirement for maintenance and will therefore be more efficient. Investigations to clarify these points conclude that selection for improved milk yield will result in increased body size, but selection for increase body size will not necessarily increase feed efficiency. Breeders should therefore select directly for yield which will result in bigger cows. Selection directly for size could result in big, inefficient and unprofitable cows.

Selection for Management Traits

Temperament and ease of milking Cows of good temperament which are easy and quickly milked are desirable under today's management systems. Nervous and excitable cows are often culled out of the herd directly at the demand of the person milking them. Alternatively cows of suspect temperament may be culled indirectly for low yield since the hormonal milk let-down reflex is impaired if the cow is unsettled.

It is desirable to improve the rate of flow of milk, but this is very difficult to measure objectively. Certainly genetic variation for this trait does exist but a faster rate of flow can be associated with a more relaxed teat canal sphincter muscle and, thus, possibly lead to greater susceptibility to udder infection. It is suggested that slow milking cows are not used to breed replacements and that sires whose progeny reports suggest an abnormally slow rate of milking be excluded from breeding programmes. More information is required before we can select positively for this trait.

Reproductive efficiency Over thirty-five per cent of all cows culled are disposed of because of reproductive disorders. Traits associated with infertility display clear, albeit low, levels of genetic variation, therefore there is scope for improvement. Cows which are persistent offenders in terms of low levels of fertility should be culled, as indeed should sires of below average fertility. Genetic studies indicate that the higher yielding cows take longer to breed, longer to conceive and require more services per conception.

In recent years it has also been confirmed that improved fertility depends on correct feeding.

Mastitis A recent MAFF survey indicates that over one million cows are diagnosed annually with clinical mastitis in the UK. This loss in production together with the high levels of culling for mastitis, leads to a drain on the financial resources of the dairy farmer. The low level of heritability (less than 10 per cent) suggests that resistance to mastitis would respond to selection but progress would be very slow.

Herd health The greatest emphasis on improving herd health should be of a managerial nature; most health traits display very low heritability, hence a dedicated effort would be required to reduce the occurrence of these traits by breeding. Progeny groups of at least 250 animals would be required in order to predict with some certainty the qualities of sires in respect of these traits.

THE IMPORTANCE OF MAINTAINING CORRECT RECORDS FOR GENETIC IMPROVEMENT

Once the correct breeding objectives have been established, it is imperative that adequate information is available for identifying the superior animals for selection and inferior animals for culling.

Production Records

Official recording schemes operate within the UK. These are administered by the various milk marketing boards who also undertake the invaluable work of processing all the raw data into meaningful indices. Milk recording provides regular information on:

1. Current daily milk yield and composition.

2. Cumulative yields and composition of current lactations to date.
3. Regular statements of completed lactations on an individual and lifetime basis.
4. Details of ancestry.
5. An important data base from which farm management and breeding (reproduction) decisions can be made.

Conformation records

Initially, type classification schemes were developed as a subjective approval of productive ability; such schemes involved the evaluation of a particular cow in relation to a 'pre-determined breed ideal'.

An enlightened development in dairy cattle breeding in the UK has ben the introduction of the 'Linear Assessment of Type.' The form of the assessment is illustrated in Figure 13.10.

The aim of the linear assessment is to describe an animal between the biological range of phenotypes that exist for a particular trait. It makes no attempt to define good or bad. Since the current idea of perfection is an opinion rather than a fact, the new scheme, when sufficient data is available, will enable us to describe in detail the physical form that nature dictates as being suitable for high productive utility over a lactation and lifetime basis.

Management Records

Each dairy farmer should keep records of reasons for culling and the lactation number in which his cows are culled. This enables herd weaknesses to be identified and rectified. Persistent offenders for mastitis, or reproductive failure can be identified and removed as can any slow milkers.

SELECTING ANIMALS OF SUPERIOR GENETIC POTENTIAL

A very important component of breeding better dairy cattle is the correct and accurate identification of those animals which are genuinely superior genetically in those characters which are to be improved.

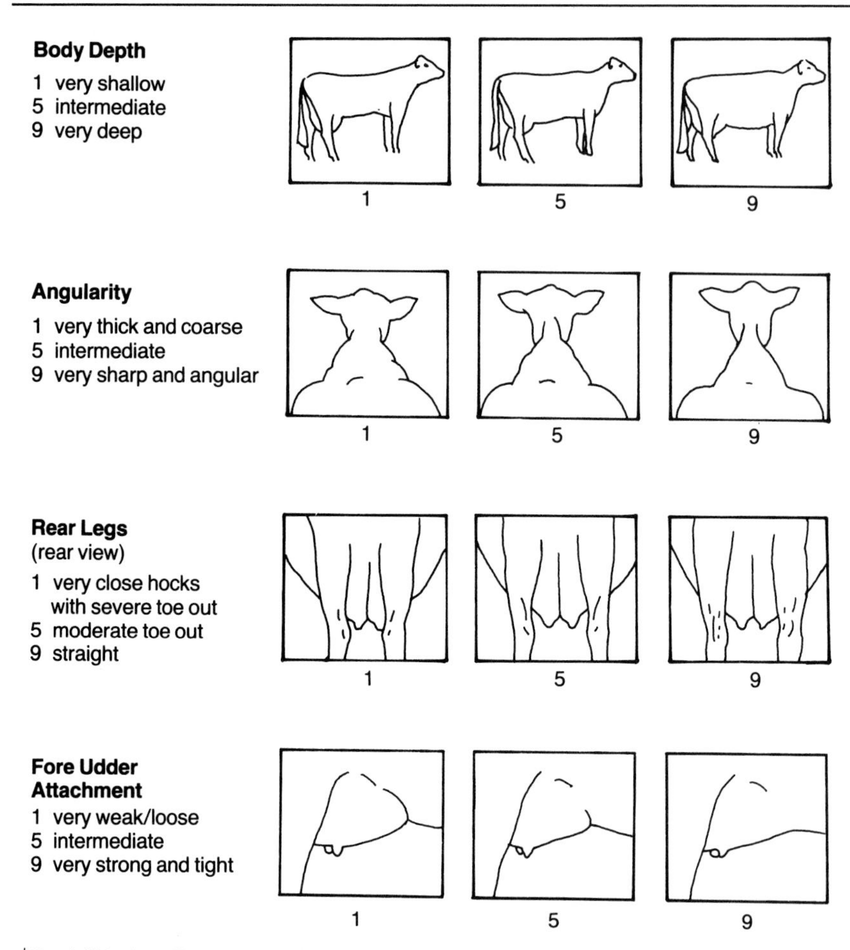

' (A total of 16 traits are illustrated in the MMB leaflet, Linear Assessment)

Figure 13.10 Examples of the range of biological phenotypes covered in the linear-type assessment
Linear Assessment, MMB leaflet

Identifying Superior Bulls

Improved contemporary comparison The only accurate method of identifying superior males is by progeny testing. The methods adopted in different countries vary but in the UK the Improved Contemporary Comparison (ICC) method of sire evaluation is used. The ICC rating of all sires which have officially recorded progeny is calculated and updated thrice annually.

An ICC rating is calculated for the following production characters based on a 305-day milk record of heifer lactations only.

- Milk yields (kgs)
- Butterfat (%)
- Butterfat (kgs)
- Protein (%)
- Protein (kgs)

The calculation can be summarised in three procedures:

1. The production record for daughters of the young bull on test are compared with the production record of daughters of other bulls within the same herd and within the same year. Comparing heifers in this way eliminates the differences in management between different herds in different years.
2. The production record will have been initially adjusted to eliminate variation due to age at first calving and season of calving.
3. An additional adjustment is made to take into account the genetic merit of the sire of the contemporary heifers and of the maternal grandsires.

Identifying Superior Cows

This is an important aspect of dairy cow breeding on both a national and a herd level. Nationally the AI companies wish to identify the very highest quality cows from which to breed sons for progeny testing. Also of increasing importance is the need to identify superior cows for embryo transfer work. Ultimately no doubt, we will have progeny evaluation methods for cows. The individual herd owner also wishes to identify his superior cows, and also his inferior cows; allowing him to decide which to breed to bulls and which to cull out of the herd.

Cow Genetic Index The more sophisticated of the two methods of identifying cows which are superior producers is the recently introduced cow genetic index (CGI). This facility however, is only available for herds participating in official milk recording schemes and which have recorded ancestry. The calculations and the assumptions on which this method is based are somewhat complex and are not described here. The CGI for a dairy cow is completed using data from four basic sources:

- The cow's own records: the production capability in lactations

1–5 for each cow is compared with those of her herd mates (i.e. contemporary comparison). The relative importance attached to the first lactation is higher than those from 2 to 5.

- The ICC rating of the sire of the cow.
- The CGI of the dam of the cow: this is based on the dam's performance over 5 lactations compared to her herdmates. It has to be noted that the CGI of the dam is partially based on information derived in turn from her sire and dam. Ultimately, therefore, the CGI for any cow is based on data from the grandparents' genetic merit as well as those of the parents.
- The herd genetic base (not to be confused with herd genetic level): this is compiled from the genetic merit of sires (ICC) and dams (CGI) of all the cows in a herd at any given time.

Cow Production Index (CPI) The facility was introduced in the mid seventies by the MMB of England and Wales as an aid for the dairy farmer in making comparisons between cows of different ages within a single herd. It cannot be usd to compare cows between herds.

The weight of fat and the weight of protein produced in the lactation are separately adjusted for age, lactation number and month of calving. This removes some of the obvious environmental factors influencing the milk record. The adjusted weights of fat and protein are combined to give the adjusted weight of total solids. The herd base is calculated by averaging all qualifying lactations for each animal in the herd. The individual cow's performance is compared with this herd base to estimate the Cow Production Index.

Example: Cow A yields 390 kgs of fat and protein in her second lactation. Adjusted for age, lactation number and month of calving this becomes 365 kgs. The herd base is 318 kgs.

$$\text{Cow A's Cow Production Index} = \frac{365}{318} \times 100 = 115$$

The CPI, which is calculated at the end of each lactation, can be used to make objective breeding decisions. The superior cows can be identified and mated to the best sires. Poor cows can be pulled out on the basis of their CPI score. If a choice is available then it can provide some guide as to which heifer calves to rear or which to retain as herd replacements or sell as surplus to requirements.

NATIONAL HERD IMPROVEMENT STRATEGY

Dairy Progeny Testing Scheme (DPTS)

Following the realisation of the importance of using the correct breeding principles to improve the national dairy herd, the Milk Marketing Board in England and Wales established a Dairy Progeny Testing Scheme. The DPTS has been running for several years and allows the practical farmer to take an active part in the breeding of better dairy cows.

Under the scheme any farmer can offer a minimum of twelve cows to be bred to young unproven bulls. Cash payments are made for each calf sired by a DPTS sire, and also for every heifer which completes its first lactation in the herd. After use, bulls are laid off until progeny test results are available. Only one bull in six on average returns to the AI stud.

The DPTS scheme provides a very useful function; without people willing to use unproven bulls, there would be no proven bulls at all.

Multiple Ovulation and Embryo Transfer (MOET)

Whereas the DPTS is a tried and tested approach which has worked well for a number of years, the Multiple Ovulation and Embryo Transfer (MOET) scheme is a theoretical, exciting but as yet untested scheme.

It utilises the principle that modern embryo transfer techniques produce litters of calves. It is, therefore, possible to assess a bull's genetic quality on the basis of his full sisters', instead of waiting for information on his progeny. This reduces the breeding cycle by half and thus increases the rate of genetic improvement. While a progeny test is a more accurate assessment of a bull's genetic merit with MOET, the accuracy lost is compensated for by this information being obtained earlier, thereby reducing the generation intervals associated with bringing new bulls and heifers back into the herd. The interesting additional feature on MOET schemes is the possibility of measuring food intake and, therefore, evaluating dairy cattle on relevant factors associated with food conversion efficiency. This was never possible with DPTS where some 40,000 progeny were being tested over many herds.

MOET Structure The structure of a typical MOET nucleus herd is shown in Figure 13.11. Essentially the 32 best cows, having been inseminated with semen from the top 8 bulls (selected on the basis of their full sister performance) are flushed to remove embryos. From these embryos it is hoped that 150 females and a similar number of males will be born and reared. The females will be milked in the nucleus herd to provide information of relevance to choose which of the bulls will be used to sire the next generation of embryos. The establishment of the MOET herd enabled a very intensive selection of the foundation females from within the top 5 per cent of females in the world. This enables a considerable 'genetic lift' in the nucleus herd, from which further improvement will be disseminated into the national herd via bulls and females surplus to the nucleus herd.

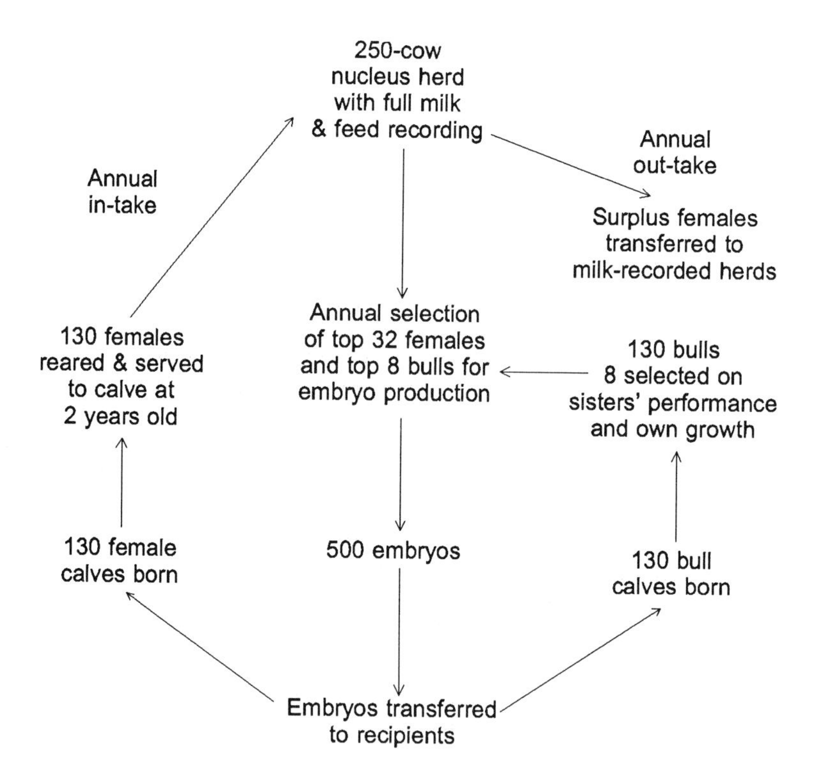

Figure 13.11 The structure of an MOET nucleus herd

TAPPING THE BEEF POTENTIAL

The primary objective of the breeder is to improve the dairy qualities of his herd. However, it should be emphasised that for many years 60 per cent of home-produced beef came from the UK dairy herd. In addition to this the dairy industry contributes dairy-cross beef animals as suckler cows in the beef herd. The precise consideration which the dairy farmer should give to beef qualities has proved emotive and confused to say the least. In recent years the trend for dairy farmers to become more specialised at the expense of the beef enterprise on the farm has resulted in selection policies which are becoming distinctly dairy-orientated. The beef industry has certainly reacted to the increased use of North American dairy cattle in our breeding programmes in recent times. In the US, beef cows outnumber dairy cows by 4 to 1, whereas in the UK, dairy cows are four times as numerous as beef cows. The future implication for the beef industry of current selection objectives in the dairy herd will be considerable.

BREEDING BETTER COWS

The main avenues which the dairy farmer may pursue to increase his returns by utilising the beef potential are:

1. Fatten off all cull cows to increase his profits and to enhance the value of the product to the beef trade.
2. Select sires whose progeny are superior for beef characteristics as well as highly superior for production ability. Beef shapes are assessed by the MMB of England and Wales in their dairy bull stud. Some 60 daughters for each sire on their DPTS are assessed to estimate a beef score for the sire. This is a useful aid to the dairy producer wishing to savour the best of both worlds.
3. Cross a proportion of the herd with beef bulls. The longer the length of life of the dairy cow in the herd the smaller will be the number of heifer replacements required. Thus an increasing proportion of the herd can be mated to beef bulls once an adequate number have been mated to a dairy bull to ensure sufficient heifer replacements.

Beef shape 1

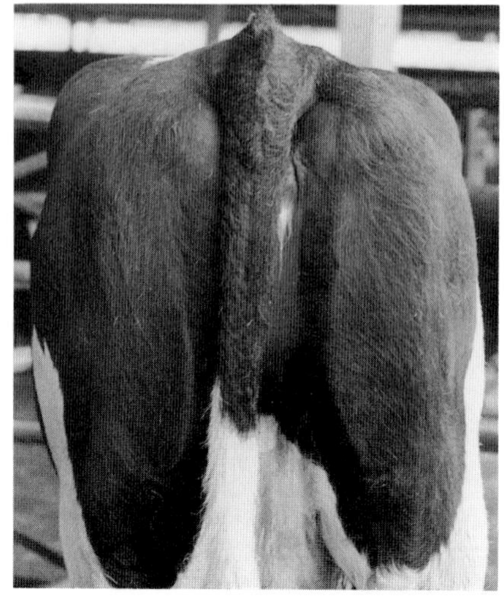

Beef shape 5

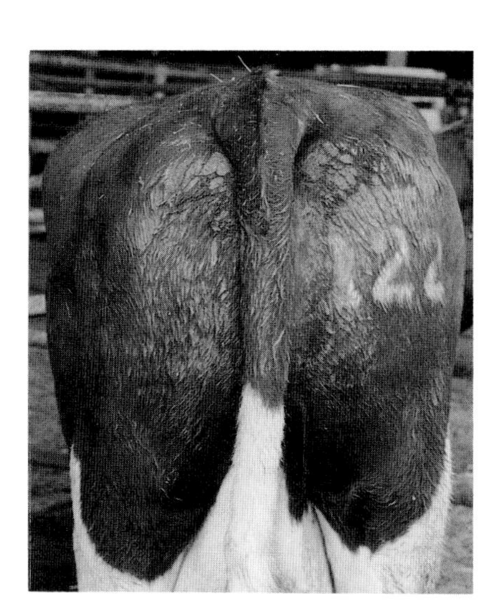

Beef Shape Assessment, MMB

Beef shape 9

4. Breeding from heifers. A large proportion of dairy farmers mate their maiden heifers to beef bulls. While this may contribute handsomely to the beef industry, it is worth thinking seriously about inseminating the heifers with the pure-bred dairy breed. This has the advantage of breeding from the very latest genetic material in the herd, so allowing more of the dairy herd itself to be mated to beef bulls and thus increasing

the selection for production within the herd. The dairy farmer may feel more confident in choosing this option for his heifers if more information regarding difficult calvings and suitability to breed was available for all progeny tested sires. While conceding that it is easier to turn a beef bull in with the heifers, there are no nutritional or management reasons why more dairy heifers should not be bred pure; calving at 2 years of age would still be desirable and possible (see Chapter 14).

CHAPTER 14

DAIRY HERD FOLLOWERS

The necessity of rearing replacements for the milking herd imposes an additional burden on the resources of all dairy farms except those carrying flying herds (i.e. where all the stock is bought in).

On the basis of an average milking life of four years per cow it is necessary to rear at least one heifer per four cows in the herd every year, which represents an appreciable cost.

There are two ways to keep this cost as low as possible. One is to increase the milking life of cows by better disease control so that fewer replacements are necessary; the other is economy in calf rearing.

Low costs in calf rearing are, in the main, achieved by restricting the quantity of whole milk consumed and using cheaper milk substitutes and calf-rearing foods. But it is false economy to carry this too far. Permanently stunted growth due to faulty nutrition will result in a lowered production potential and a loss of revenue when the animal is in milk that will far outweigh the saving made at rearing time.

Dairy heifers must be reared on the modern dairy farm with a view to the age at which they are required to calve. It is very much a matter of opinion as to what a farmer may wish to do but the facts are simply that, in the case of Friesian calves, with a birthweight of 40 kg, the requirement is to increase weight to 510 kg immediately prior to calving. This gives a target of about 0.64 kg per day liveweight increase for a two-year calving heifer. If calving is extended to a 2½-year period, then the growth rate is less. This liveweight gain is not necessarily attained at the same rate all through the growing period. In early life, the first three months, the requirement is 0.5 kg/day increase; in the period 12–18 months the rate is 0.7 kg/day. The simple fact is that heifers must be big enough to serve (330 kg), at the right age, nine months before calving.

Table 14.1 Target Liveweights over the Life of a Friesian Heifer

Age	*Weight kg*	*Liveweight gain kg per day*	*Weight band measurement cm*
Average birthweight	40		
5 weeks	55	0.5	83
3 months	85	0.5	98
6 months	140	0.6	118
12 months	270	0.7	150
Service	330	0.7	161
18 months	370	0.5	167
Just before calving	510	0.7	185
Just after calving	450		

Source: *Rearing Friesian Dairy Heifers*, MAFF.

High liveweight gain during the winter period can be uneconomic, bearing in mind that there is compensatory growth on grass during the early summer. The problem is usually that many heifers are left out on an off-lying field and cared for less than they should be. When summer grass deteriorates the heifers will not continue to grow unless they receive supplementary feed. On lowland farms the need is to conserve land for the milking cows, so it is desirable to have a two-year calving pattern. On farms with areas of marginal land such as banks and slopes, heifer rearing is not using potential cow pasture so calving is more likely to be at two and a half years.

The other factor to be considered is when the heifers are required to enter the herd as down calvers. Some herds are block calving, so heifers have to conform to a strict calving pattern of two years. Other herds have spread calving patterns; there is no pressure to calve early, except that land should be economically used and any extension beyond two and a half years' calving is likely to be uneconomic.

Table 14.2 shows the number of replacements to be reared on a farm where calving is at 2 years rather than 3 years. The additional replacements require extra land, capital and, probably, buildings to be tied up.

There are two occasions when it could possibly be admissable for heifers to calve after 2 years and even up to 3.

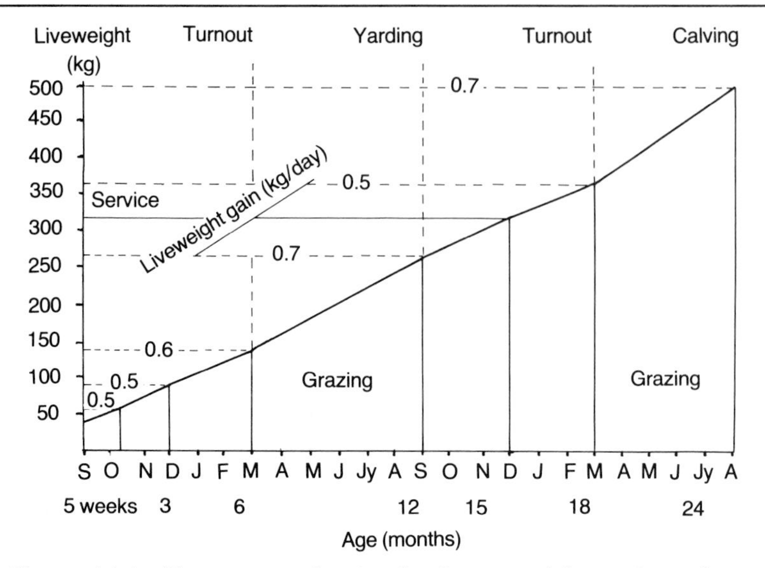

Figure 14.1 Target growth rates for 2-year calving autumn-born Friesian heifers

Rearing Friesian Dairy Heifers, MAFF

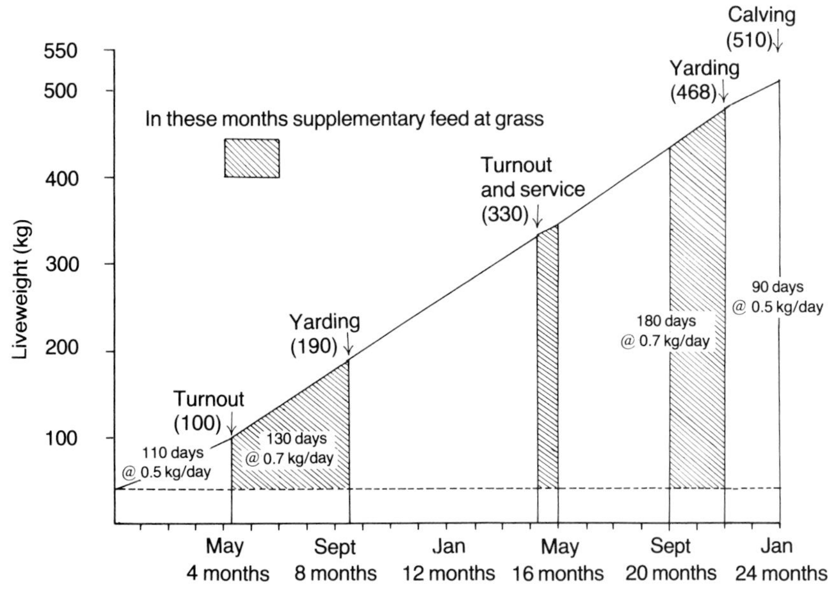

Figure 14.2 Target growth rates for 2-year calving spring-born Friesian heifers

Rearing Friesian Dairy Heifers, MAFF

(i) Where there is a considerable surplus of grazing/conservation area which cannot be used by the dairy herd. This may occur on a farm which has a rather low quota, meaning that no more cows could be milked without producing over-quota milk.

Rearing costs would be extremely low and the resultant value of heifer sales could produce a nest egg which would be used for capital projects, even the purchase of quota.

(ii) In the case of pedigree sales, heifers are bought on the basis of their parentage. Even more money will be paid if the animal is well grown. A move from 2 to 2½ years calving will certainly produce larger heifers and could additionally alow them to reach higher levels of yield. These factors could combine in a pedigree situation to realise much higher prices than might normally be the case.

Once a target calving date and weight have been decided, it is then very important to give the calf a good start in life; the cow's first milk is essential.

Table 14.2 Number of Replacements to be Reared

Age at first calving	*2 years*	*2½ years*	*3 years*
0–12 months old	25	25	25
12–24 months old	25	25	25
24–36 months old	—	13	25
Total no. of replacements on farm during any given year	50	63	75

Source: *Rearing Friesian Dairy Heifers*, MAFF.

VALUE OF COLOSTRUM

Unless calves are reared so that they achieve full growth with freedom from disease, the best-planned breeding programme will fail.

The colostrum secreted by the cow during the first four days when she is in milk is of special significance in calf nutrition. Recent research has shown that it is absolutely vital that the calf should receive colostrum in the first six- to twelve-hour period.

Colostrum has the following properties:

- It contains extremely valuable antibodies which give the calf an immediate resistance to bacterial diseases, such as white scour. The antibodies can only be passed through the abomasal wall undigested in the first 24 hours and possibly only during the first 6 hours.
- It is highly digestible, having a nutritive value approximately 40 per cent greater than ordinary milk, giving a soft curd during the process of digestion.
- It is laxative in effect, and so helps in the expulsion of the foetal dung. The passage of this black jelly-like faecal matter usually indicates that the calf has suckled for the first time.
- Provided the dam has received adequate green food in the form of silage, kale, or dried grass, the colostrum is rich in carotene (the precursor of Vitamin A, which is important for disease resistance).

SUBSTITUTES FOR COLOSTRUM

Whenever possible newly-born calves should have colostrum; there is no adequate substitute for it.

Calves taken away from their dams at birth should be bucket-fed on their dams' milk for at least three or four days. And if a cow dies at calving or if she has been milked for longer than four days before calving, an effort should be made to obtain colostrum for her calf within six hours of birth. This can often be done by keeping the cow's colostrum in a refrigerator until the calf is born or by using surplus colostrum from another freshly-calved cow or from a cow that is being milked before calving and is still giving colostrum.

Where colostrum is not available from any source, then, as a last resort, a substitute can be used. A well-known recipe is:

> Whip up a fresh egg in 0.85 litres of milk; add 0.28 litres of warm water, 1 teaspoonful cod liver oil and 1 dessert-spoonful castor oil. The quantities given will make one feed; give three feeds per day for at least four days. Once the calf had defecated normally, omit the castor oil from the mixture.

METHODS OF CALF REARING

Work by Blaxter has shown that a young calf has a high efficiency in using milk for growth and furthermore, when young it will make more rapid growth in proportion to the amount of food fed than it will at a later stage in life.

Therefore, restriction of milk or milk substitute at an early age can lead to stunted growth which can only be corrected later on by heavier and more expensive feeding.

This again points to the need for care in calf feeding so that economy is not carried too far.

Against this background the common methods of calf rearing can be outlined and discussed.

Nurse Cow Rearing

In this system the calves are left with their dams for the colostral period of four days and are then transferred to a nurse cow capable of suckling up to four calves at a time according to her milk yield.

Excellent calves can be reared by this method. It saves labour and, compared with pail feeding, it reduces calf mortality. It is the customary method of rearing bull calves for service.

Its main disadvantage is that it tends to be expensive if the value of milk is taken into account. Secondly, nurse cow rearing may predispose the calf to cross-suckling later in life, which is very undesirable. But, where suitable nurse cows are available—cows that are to be culled from the milking herd for reasons other than disease—it has much to recommend it.

Suckled calves usually develop more quickly than those reared by any other method.

The number of calves a nurse cow will rear must be adjusted to the yield of the cow so that each calf gets about 450 litres of milk over a period of twelve to sixteen weeks. Suckling three times a day is recommended for the first week; thereafter twice a day is satisfactory.

Calves can be weaned as soon as they are consuming adequate amounts of other foods, say, 2.3 kg of hay and 1.4 kg of concentrates a day. Then fresh young calves can be put to the cow, the number adjusted according to her declining yield.

It is, of course, essential to feed nurse cows on the same lines as those in the dairy herd so that their maintenance and production needs are fully met.

The Early Weaning System

The traditional method of feeding the calf with whole milk has been increasingly replaced by the 'early weaning' system. This change took place largely because the cost of milk substitute is almost half the price of whole milk and as such is a much more economic way of rearing calves.

In the early weaning system, ideally, no milk is fed after the colostral period of four days. Over the next five weeks, milk equivalent feeds are provided followed by weaning on to a ration wholly made up of dry feeds plus water. The system is described in detail in Table 14.3.

This system, by reducing the period of bucket feeding to five weeks, saves labour. And it saves accommodation by reducing the need for individual penning of the calves to five rather than twelve weeks—once weaned, calves can be penned communally according to age. By early consumption of dry foods, rearing costs are also reduced and early development of the rumen is encouraged.

It should be clearly understood that success in early weaning depends on using good hay/straw to encourage rumination, and using concentrates of the highest feeding value to ensure an adequate nutrient intake by the calves.

This system is ideal for spring-term calves (i.e. March-May), where considerable savings can be made by rearing them outside on grass. A straw bale hut can be used to provide cheap shelter.

A clean pasture (a new ley not previously grazed by cattle) should be chosen and rotational paddock system of grazing adopted, allowing a total area of 0.13 hectare per calf. Grazing allows the hay and concentrate consumption to be reduced to not more than 1.4 kg concentrates and 1.4 kg hay per calf per day.

The Ad Lib Feeding System

The cold ad lib acidified feeding system is rapidly gaining popularity. The milk powder is acidified so that the made-up liquid food will be preserved for a period of three days or more without going sour. This of course can be affected by weather conditions.

Table 14.3 The Early Weaning System

1. Feeding during the first five weeks of life

1–4 days colostrum fed as described in text.
A standard mix for milk equivalent is 125 g/litre for 2 × feeding.

5–10 days

	whole milk or milk equivalent
Jersey	1.5 litre
Ayrshire and Guernsey	2.25 litre
Friesian	3.0 litre

11–28 days

	milk equivalent
Jersey	3.0 litre
Ayrshire and Guernsey	3.5 litre
Friesian	4.5 litre

29–35 days

Reduce milk equivalent by approximately 0.57 litre per day until calf is weaned on 35th day.

Remarks

Early weaning concentrate mixture (see below) with *good* hay/straw and water available ad lib from the fifth day.

2. Feeding from sixth to twelfth week

Early weaning concentrate mixture plus good hay/straw and water ad lib continues to be fed until consumption reaches the following levels:

Jersey	1.8 kg
Ayrshire and Guernsey	2.0 kg
Friesian	2.5 kg

Remarks

(a) Proprietary 'early weaning' pellets (16–18 per cent CP) can be purchased, or the following 'early weaning' concentrate can be made up:

	per cent
Flaked maize	50
Rolled oats or barley	20
Linseed cake	10
Dried skimmed milk	10
White fishmeal	10

Plus ½–1 per cent of synthetic A and D vitamin supplement.

(b) Good hay or barley straw must be available.

(c) After the 12th–14th week the early weaning concentrates can be replaced by a concentrate mixture as fed under the 'traditional' system, or proprietary calf-rearing pencils can be used.

The acidified powder is mixed with water at the rate of 1.4 kg per ten litres. In cold weather it is advisable to take the chill off the water before mixing.

The acidified milk substitute is fed from teats which run on a pipeline from drums of 150–200 litres capacity. It is usual to allow one teat per six calves. The drums should be provided with lids.

One of the attractions of the system is the grouping of calves and avoidance of cleaning individual pens and feeding equipment.

Group pens should have rendered walls where calves are in contact with them and, since more liquid is consumed and voided by calves, a concrete floor with a good fall and drainage. The average daily intake would be eight litres of acidified milk substitute.

FEEDING RECOMMENDATIONS

First 4 days—Colostrum.
5–7 days—50:50 colostrum/milk:acidified milk substitute.
8 days—Acidified milk substitute.

Calves should be fed on acidified milk substitute ad lib for about three weeks before being weaned on a gradual basis. It needs to be gradual because of the lack of rumen development at this stage, so it should take place over a two-week period. The dry foods, hay and weaning pellets should be available and their intake will increase as the acidified milk substitute is reduced.

The system is successful and popular because:

- It is a simple system to operate.
- There is flexibility of labour—particularly at weekends.
- The calves seem to grow well and show bloom.
- The capital costs are minimal.

The cold ad lib system has declined in popularity because there is still quite a high labour input in mixing large quantities of liquid and the cost is greater than for conventional bucket rearing.

In contrast, the machine dispenser requires only milk powder to be put into the hopper and to be given a daily clean.

The dispenser is popular, especially on large dairy farms, to take the beef cross calves and bull calves on for 2–3 weeks before sale. They increase in value far beyond their feed cost and sell very well.

CALF MANAGEMENT

Certain points in the general management of calves reared on the bucket require special emphasis.

The newly-born calf should be housed in a warm box or pen, well ventilated, but free from draughts; it should be rubbed well with wisps of hay or straw to prevent chilling and the navel should be dressed, before it has dried, with tincture of iodine or powdered bluestone (copper sulphate) to prevent any bacterial infection.

The calf should not be over-fed with milk at any one feed, since in nature a calf suckles at least three times daily. Too large a feed of milk means that the amount which is surplus to the capacity of the fourth stomach spills into the rumen (first stomach), ferments and rapidly causes bloat.

Milk fed below blood heat may lead to similar trouble except when available ad lib.

Use sterilised feeding buckets and try to encourage the calves to drink as slowly as possible. Feeding buckets with rubber teats have many adherents because they force the calf to swallow slowly, as in suckling. This method reduces the incidence of digestive troubles.

With purchased calves, which have probably suffered a period of starvation, give a first feed of 1.2 litres of a warm (38°C) 5 per cent solution of glucose, that is, 50–100 g in 1–2 litres of water, according to the size of the calf. This helps to satisfy hunger without the risk of digestive scour—milk can then be fed at the next meal.

To combat infectious scours due to *Escherichia coli* infection, a prophylaxis of a proprietary *E. coli* serum can be given by a veterinary surgeon; otherwise, antibiotics such as terramycin or aureomycin can be used under veterinary supervision.

To teach a calf to drink, back it into a corner of its box or pen, with its milk feed in a shallow bowl or small bucket. To make it suck insert one finger of the right hand into the calf's mouth and then guide its head so that its mouth is immersed in the milk, holding the bowl or bucket in the left hand.

The calf sucks readily if its head is slightly raised, so at first do not attempt to make it drink with its head down. As the calf sucks and swallows the milk, slowly withdraw the finger so that the calf begins to drink rather than suck.

After the second or third feed, a calf will usually drink from the bucket at head level, without further trouble. Patience is essential.

Feeding Individually

If calves are housed in groups then feed them individually. Use individual feeding yokes, or tie up each calf with a leather strap at feeding time; and do not release them until their lips have dried, as this will reduce the risk of cross-suckling.

Watch the calves carefully as loss of appetite is a danger signal. It may mean the onset of pneumonia or white scour which are both highly infectious diseases. It may also be the first sign of digestive scour.

Make all changes in the calves' diet as slowly as possible. This principle is indicated in the feeding scheme outlined earlier in this chapter.

Housing for calves need not be elaborate. The main objectives are to keep the young calves warm, provide ample bedding, and to construct interior wall surfaces and floors so that they can easily be cleaned and disinfected. Wherever possible ample sunlight should be admitted.

After each batch of calves are weaned, the calf-house should be given a thorough cleaning, using 5 per cent proprietary detergent steriliser to remove dirt adhering on walls and fittings. Following this with a short rest of five to seven days is also very beneficial in preventing a build-up of infection in the premises, which can cause severe mortality from scouring or calf diphtheria in buildings where rearing is a continuous process. A power wash system is advisable.

Disbudding

Under the Protection of Animals (Anaesthetics) Act of 1964, it is illegal to disbud a calf over seven days old except under an anaesthetic. The standard dehorning method is with the hot iron.

Electrically heated or gas-heated dehorning irons are available. The anaesthetic should be injected to each horn bud at least 15 minutes before using the iron. (The site of injection is critical and it is essential that operators receive qualified instruction).

Disbudding is best carried out at 2–3 weeks of age. The operation becomes progressively harder with increasing age.

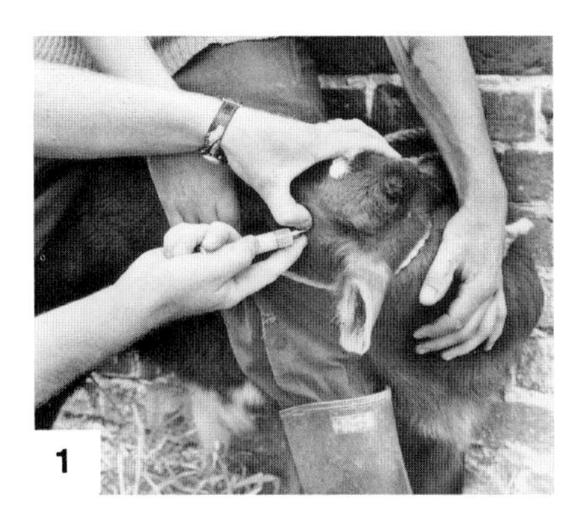

1

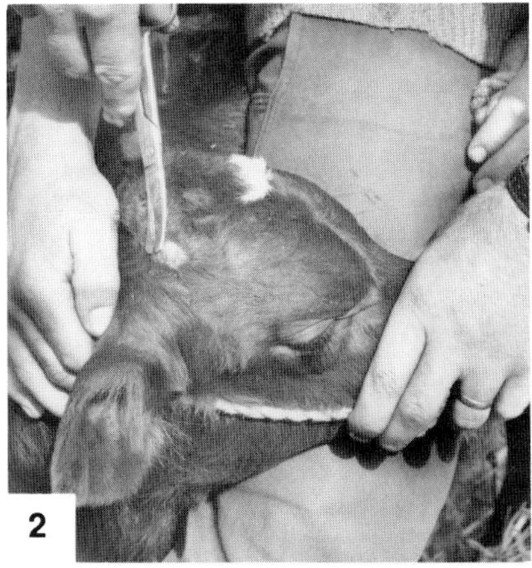

2

Dehorning a Calf

1 With assistant holding calf in firm grip, feel for groove in skull between eye and horn bud. This contains nerves which run to horn bud and an injection of 4 cc of anaesthetic will 'freeze' them. Hold needle flat against skull and guide point in (away from eye) with your spare thumb. Withdraw plunger slightly and check for blood. If blood is present, withdraw the needle and inject again in a slightly different position, adjacent to previous insertion. Repeat on other side of head.

2 Allow at least 15 minutes for the anaesthetic to take full effect. During this time, clip back hair around horns to expose immature buds. Note how head is held in a bent-neck grip around the thigh.

3 Having heated the dehorning iron (see that it gets red hot), apply it squarely, ensuring that bud is fully enclosed in burning ring. Press down iron and rotate it uniformly to speed burning action and shift any build-up of burned material. Burn down until layer of white cartilage is reached; then remove bud with a knife or a flick with iron.

4 A correctly performed operation will leave two clean bud sockets—the skin and skull having been burned but no flesh. Slight bleeding can be stopped by cauterising. The wound can be dusted with sulphanilamide powder.

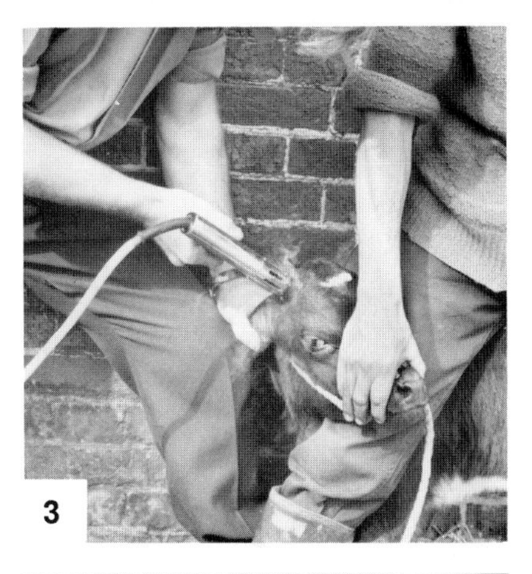

3

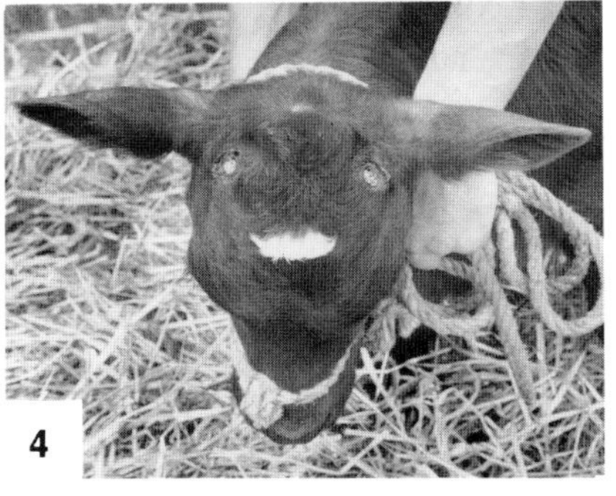

4

Ear Marking

Calves are ear-marked in the right ear for registration purposes. This is usually done by tattooing. Care should be taken to avoid tattooing across the ear veins, thus avoiding septic ears or indistinct tattoo marks owing to incomplete closure of the forceps. For Friesians a red marking ink shows up best. Ear tags are now approved for ear-marking under the Tuberculosis Order 1964 by the MAFF.

Removal of Extra Teats

When calves are ear-marked it is a good plan to examine them for any extra teats on the udder. At this age, their removal can be easily effected with a pair of sharp curved surgical scissors without the risk of damaging the udder tissues.

REARING BULL CALVES

As previously noted, bull calves are often reared on the suckling system. There is no reason, however, why pail feeding should not be equally effective, provided the level of feeding is raised some 20–25 per cent higher than with heifer calves. For example, where a heifer calf receives 4.5 litres of whole milk per day, a bull calf should have 5.5 litres. Likewise, where a heifer calf would be fed up to a maximum of 1.8 kg of concentrates, a bull calf should get 2.3kg.

Feeding a bull calf generously encourages quick growth and enables the animal to be put to service early in life, thus giving an earlier progeny test.

A bull calf may be capable of service before ten months of age, and should consequently be segregated from heifers after weaning.

Castration

Bull calves which are raised for beef are usually castrated before twelve weeks of age.

Since January 1, 1983 only a veterinary surgeon, using anaesthetic, has been permitted to castrate any farm animal over the age of 2 months. The use without anaesthetic of rubber rings or other devices to constrict the flow of blood to the scrotum is still allowed, but only during the first week of an animals life. The ring is applied to the scrotum using a pair of special pliers. Once the operator is sure that the immature testicles are through the ring, the pliers are closed, allowing the ring to be eased off.

A method of castration which is much favoured is the Burdizzo Bloodless Castrator. Young calves can be castrated while standing. Very young calves, under a month old, are laid down, on their side.

The operator's assistant should sit on the calf's head and keep the uppermost hind leg of the calf pulled well forward. Before closing the jaws of the castrator make sure that the spermatic cord is between them, and make two independent closures for each cord.

Castration with a knife is very effective and probably more reliable than the bloodless castrator, but is a job for the veterinary surgeon.

Feeding after Weaning

It is wise to continue to give from 1.0 to 2.0 kg of concentrates per day from weaning until the calves are eight to nine months old, and relate concentrate feeding to liveweight gain. Their increasing appetite for food can be met by feeding extra hay, straw or silage.

The equipment needed for ad lib cold milk feeding is cheap and simple. Here the calves are sucking through a rack of teats and siphon tubes from insulated containers made out of plastic dustbins.

Use sterilised feeding buckets and try to encourage the calves to drink as slowly as possible.

During this period growing calves will consume from 0.30 kg to 0.45 kg of hay per month of age each day. Thus a six-month-old Jersey calf should get about 2.0 kg of hay or 6.0 kg of silage, whereas a Friesian calf of similar age should get 3.0 kg of hay or 9.0 kg of silage—in fact bulk feeds are fed ad lib.

Attention must be given to the calves attaining the target liveweight applicable to their age. If the target is not being reached then the food input must be increased until the deficit is corrected. Care must be taken to ensure that when young stock are housed away from the steading, they are observed regularly and a check is made on their progress. A visual appraisal is of course useful but the only sure way to know that cattle are gaining weight at the planned rate is to weigh them regularly. Once a month would be ideal.

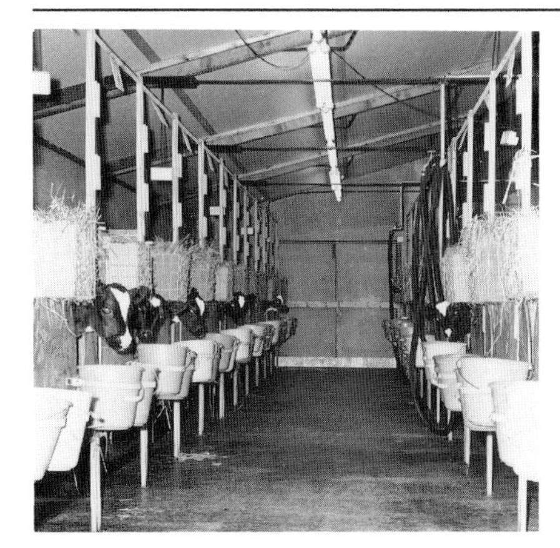

(Left)
A calf-rearing shed at the National Institute for Research in Dairying.

(Below)
Calf cubicles at the Somerset Farm Institute.

YOUNG STOCK AT GRASS

Calves are usually turned out to grass at about six months of age, by which time they are sufficiently well grown to make the best use of it. Grazing is attendant with the risk of husk and worm infection, usually more manifest in the autumn months; at the first sign of coughing, a feed of hay should be given at night and if possible the animal should be housed in a shed or yard to prevent a serious onset of husk.

Experiments in New Zealand with identical twins have shown that calves rotationally grazed on new leys outweigh their twins, grazed on low-quality pasture, by over 50 kg at ten months old. The old system of always using a paddock convenient to the buildings as a permanent calf paddock—a common practice on many farms in Britain—cannot be too strongly condemned.

Provided clean grazing is available, calves can be turned out to grass by early April with 1.0–2.0 kg of concentrates per head to maintain nutrient intake for the first three to four weeks. Choose a sheltered field for this purpose. To minimise risk of husk infection, a vaccine can be given in two doses one month apart, starting one month before turnout. This treatment is expensive, but is recommended where the incidence of husk has been high in the past.

On farms where losses from blackleg have occurred, preventive vaccination of all young calves at turning-out time should be carried out.

Cold milk feeding of calves at pasture. The feeding apparatus is positioned to offer shelter to suckling calves. Trials at the Grassland Research Institute at Hurley have shown that the health of calves fed on cold milk is in no way inferior to calves fed on warm milk.

AGE FOR BULLING

The age for bulling depends upon the breed and also upon the requirement of the desired calving pattern. Certainly the smaller breeds tend to become sexually mature earlier than the larger breeds, but this has to be correlated to the practical demands of the annual milk production cycle.

It is generally suggested that heifers should not calve before twenty-three months old but should calve as soon as possible after that age. In an autumn-calving herd the calving of heifers should be arranged so that animals born in August could calve at twenty-five months old, whereas any animals born in November and December need to be pushed to calve at twenty-three months old, thus preserving the calving pattern.

The majority of herds have calving intervals (the average time interval between successive calves born to individual cows) of considerably over 365 days. If heifers are not brought in at the right time then the calving pattern of the whole herd is likely to rapidly become dispersed over the whole of the winter period.

Some degree of block calving should be aimed for as this does simplify certain aspects of herd management.

A Basic Ration for Bulling Heifers

For a bulling heifer, depending on the breed, the basic ration in winter should be at least 0.9 kg per day concentrated foods, as fed for milk production, or its equivalent in the form of mineralised barley with protein supplement, together with hay/straw at 0.2–0.4 kg per day per month of age. Thus for a Jersey heifer, fifteen months old, feed:

- 3.4 kg hay (or hay equivalent in silage)
- 1.0–2.5 kg concentrates
- plus straw to appetite

And for a Friesian heifer, twenty months old, feed:

- 7 kg of hay (or hay equivalent in silage)
- 1.0–2.5 kg of concentrates
- plus straw to appetite

These rations would have to be adjusted according to the rate of liveweight gain, since it is undesirable to have heifers in too good a condition in the early stages of pregnancy.

Feeding the In-Calf Heifer

During pregnancy the aim should be to steadily increase condition from the relative leanness of the bulling heifer to the well-conditioned down-calver.

From the fifth month of pregnancy onwards, the demands of the developing foetus and udder rapidly increase. With autumn-calving heifers, this demand on bodily reserves is usually adequately met from the grazing. Spring-calving heifers, however, should be given a winter diet at least thirty per cent greater than their actual maintenance requirements.

Table 14.4 can be taken as a guide. The full hay equivalent could be provided in silage. Where the bulk feed is not of good enough quality, then it may be necessary to feed some concentrates to ensure that the heifers attain their target calving weight.

Table 14.4 Winter Ration for In-calf Heifers (all in kilograms)

Breed	*Liveweight*	*Basic ration hay*	*Supplemented by silage*	*or kale*
Jersey	317	4.5	9.5	12.7
Ayrshire/Guernsey	381	5.4	9.5	12.7
Friesian	509	8.2	9.5	12.7

The liveweight gain after the animal has been confirmed in calf should be 0.66 kg per day, based on the following calculation: deducting bulling weight (330 kg) from calving weight (510 kg) gives a gain of 180 kg to be made in 270 days, or 0.66 kg per day.

When in-calf heifers are out to grass and sufficient grass is available, this should be adequate to maintain the required growth rate through the summer. If the grass becomes scarce, then immediate steps should be taken to supplement the ration, so that the animals continue to gain weight as required.

Animals must be brought to calving in a fit but not fat condition.

Use of Urea Supplements The use of urea as a supplement to barley or oat straw fed ad lib to dairy heifers is now increasingly common. In block form or as liquid formulations, the urea is consumed in small doses over twenty-four hours, thus reducing the risk of excess ammonia being produced in the rumen. Appetite for straw is raised by some forty per cent, but rationing of the straw is not recommended; when fed ad lib the cattle select the most palatable roughage, the residue being used as litter.

Urea supplements, though expensive, do offer the opportunity to use barley straw in place of hay. This reduces the area of grass required to rear replacement heifers, enabling more feed grain to be grown, resulting in greater profit to the farm as a whole and increased productivity by utilisation of arable by-products. Urea is used by the farmer as a pre-mix, in conjunction with mineral and vitamin supplements. It is important to feed barley or other cereal feeds with the pre-mix since the added energy is necessary for efficient protein break down and utilisation.

ECONOMICS OF REARING REPLACEMENTS

In economic terms it is difficult to justify the rearing of dairy heifers. The details are shown in Table 14.5.

Table 14.5 Friesian Dairy Followers (per Heifer reared)

Performance Level	*Low* £	*Average* £	*High* £
Value of heifer (allowing for culls)	590	590	590
Less value of calf (allowing for mortality)	80	80	80
OUTPUT	510	510	510
Variable costs:			
Concentrate Costs	150	130	110
Miscellaneous Variable Costs	40	40	40
TOTAL VARIABLE COSTS (excluding forage)	190	170	150
GROSS MARGIN per Heifer, before deducting Forage Variable Costs	320	340	360
Forage Variable Costs (fert. and seed)	59	65	68
GROSS MARGIN per Heifer	261	275	292
Forage Hectares (Acres) per Heifer reared	0.95 (2.3)	0.725 (1.8)	0.575 (1.4)
GROSS MARGIN per Forage Hectare	275	380	508
GROSS MARGIN per Forage Acre	111	154	206

Source: Farm Management Pocketbook, J. Nix, Wye College, 1989.

Notes:
1. The price for the heifers is based on the selling price of down-calving heifers, after due allowance for culls.
2. Costs in this report suggest that the high performance animals use less concentrate by being more efficient with feed use and having a lower average calving age.

3. The variable costs include the cost of straw, about 1 tonne being used on average, although this varies with the size of the building and extent of wintering outside.
4. The high performance animals are reared on a more intensive stocking rate and achieve a lower average calving age.
5. The gross margin per hectare virtually doubles from the low to the high performance level. This spells out in strict financial terms that good intensive husbandry makes money.

If dairy heifers are not likely to be reared well, with attention being paid to detail, then it is better to cease rearing and purchase heifers on the open market.

Although home-reared heifers may be difficult to justify economically, it may be worthwhile in the following circumstances:

- Where there is a well-planned breeding policy based on progeny-tested sires allowing the genetic potential of the herd to be improved. If this genetic potential is linked to high levels of production, then obviously an expanding market for surplus stock would exist. The reward for the exercise would be in stock sales at a premium rather than in terms of margin per acre.
- Where land not suitable for the dairy herd is available for summer grazing.
- Where wintering can be cheapened by the feeding of straw or arable by-products, or by labour-saving systems, e.g. self-feeding of silage utilising maximum bulk of feeds rather than early-cut quality silage.

  A self-contained herd is, it should be noted, more favourably placed than flying herds to eliminate such diseases as brucellosis and mastitis.

  For home-bred replacements to be reasonably economic there are two very important factors to be borne in mind:

1. Heifers should calve as young as possible. Calving at two years of age means that less land needs to be allocated to the rearing enterprise, leaving more to be used by other farm enterprises with higher gross margins. Rearing livestock is a means of investing capital, which in the case of a young farmer starting a dairy business is not perhaps the best use of capital, particularly where there might be cash flow problems. Three-year calving will certainly tie up capital for a longer period than a two-year calving policy.

2. Where young stock are to be reared it is important to follow an intensive rearing programme and this policy must include the move towards a tighter stocking rate. This can be achieved by intensive grazing systems during the summer months. Paddocks provide a suitable degree of control, and can put heifer rearing into a much more profitable situation.

CONTROLLED BREEDING—PROSTAGLANDIN

In recent years there has been an increasing interest in the technique of breeding heifers to fit into a planned calving pattern.

This involves the use of prostaglandin, a naturally produced substance in the body which can be artificially produced and used for controlled breeding.

In a normal cow reproductive cycle, a corpus luteum is formed which will be retained if a pregnancy is established (or sometimes as a physiological malfunction). An injection of prostaglandin will remove the corpus luteum and allows oestrus to occur, usually within 7 days.

If used on a number of cows with functional corpora lutea then synchronisation of oestrus is possible. Cows can then be organised to fit into a planned calving pattern. This technique, however, is no substitute for good husbandry and breeding practice. With heifers, prostaglandin can be used in a two-injection system. The first injection is followed by a second one 11 days later. The heifers must be inseminated or served at 72 and 96 hours after the second injection. Service may also be carried out if oestrus is observed after the first injection.

A simple timetable for the two injection system would be: 1st injection—eleven days later 2nd injection at 10am—serve three days later at 10am and next day at 10am. Heifers must be well grown and in good condition for successful mating. If this is not the case, good results even with drugs are by no means assured. It is also very important that the animals are cycling regularly for success to be achieved.

BRUCELLOSIS ERADICATION SCHEME

Brucellosis in cattle is a contagious disease which may cause abortion; in many cases it causes infertility and it certainly leads to reduced milk yields.

The Brucellosis (Accredited Herds) Scheme was initiated in 1967 by the Ministry of Agriculture. Eventually the intention was to establish brucellosis-free herds over the country. Eradication of brucellosis commenced in November 1971 on a voluntary basis, but in three main areas of Britain it was made compulsory by November 1972, and gradually other areas were included in the compulsory scheme. The last eradication areas in Scotland and England entered the compulsory scheme in November 1979.

To reach accredited status a herd was obliged to pass three consecutive blood tests. Accredited herds are tested periodically by the Ministry of Agriculture to ensure that they remain free from the disease. Dairy herds are also tested by the Marketing Boards on a monthly basis and any failures are communicated to the Ministry. Abortions at any time also have to be communicated to the Ministry. Where animals are found to be infected, there is a compulsory slaughter policy; compensation is paid by the Ministry.

The eradication of brucellosis have proceeded with considerable success to the point where few cases occur at the present time.

Although S19 vaccine was used at one time, this is no longer allowed.

HOUSING AND MANAGING BULLS

Many farmers have traditionally reared dairy heifers on their farms. Once bitten by the 'pedigree breeding bug' then the next step forward is to rear a bull for use on the farm.

Rearing bulls has not been very popular since the advent of AI. However after more than fifty years of the AI service, people have become much more knowledgeable and appreciative of good breeding potential, resulting in bull breeding becoming popular again. This trend has certainly been encouraged by the abolition of the Bull Licensing Scheme during the mid-eighties.

Acquiring or breeding a bull is certainly within the means of many farmers, but nonetheless keeping a bull is no simple matter. It takes several years for a bull to prove itself with respect to its progeny. All the advantages of progeny testing will be lost, however, if by the time the test is completed—say by five and half years of age at the earliest—the bull concerned is dead. Yet this is often the case.

On many farms the early disposal of a bull is due primarily to the lack of suitable housing. As bulls get older their tempers usually deteriorate and they become more difficult to handle unless controlled by specially designed buildings and apparatus.

Good Housing

Proper housing for stock bulls is therefore essential to any long-term breeding programme, as is a policy of good bull management. The temperament of a bull is firstly conditioned by heredity, but even a quiet animal can be ruined by wrong methods of handling.

The essentials of good bull accommodation are:

1. A strongly-built loose-box, minimum dimensions 3 m × 3 m with smooth inside walls, opening on to an exercise yard of double that area. The feeding manger in the box should be placed so that the bull can be fed from outside the box.
2. There should be a stout cable running overhead, from inside the box, above the feeding manger and along the full length of the yard, to which the bull is attached by a chain (with two or three swivel links) which passes through his ring and around his horns.

   In polled breeds a head collar can be fitted to which the chain can be attached at the top of the bull's head, mid-way between the ears.

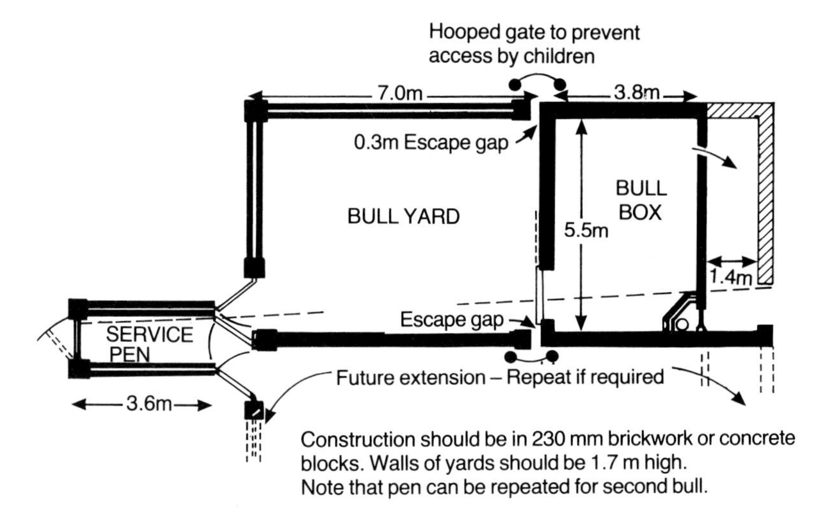

Figure 14.3 Plan for a bull pen

The length of the tethering chain is important. It should equal the height of the overhead wire above ground level; it will not then prevent the bull from lying down nor foul his forelegs when he does lie down.

Secured in this manner the bull is perfectly free to take exercise but is efficiently under control and can be caught and secured without the stockman entering the yard or box.

3. Built in to the yard should be a service crate to which a cow can be admitted for service. A gate normally shutting off the crate from the exercise yard is then opened to admit the bull without releasing him from his overhead tether.

   Never, ever, take liberties with a bull; they are always a potential hazard.

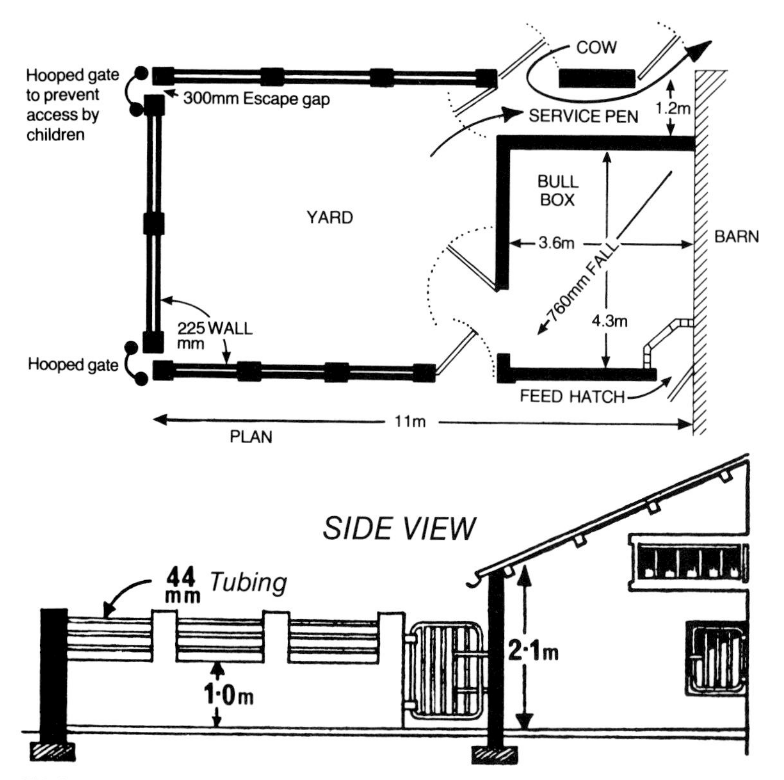

This bull pen layout allows most routine jobs such as feeding and mating to be seen to without going into the pen. Main structure is 230 mm brickwork or concrete blocks and roof of asbestos cement. Safety measures must be built in for the operator. If in doubt approach the Health and Safety Executive.

Figure 14.4 Alternative bull housing

Siting the Bull Pen

With old and heavy bulls, it is often considered necessary to use a service ramp to take the weight off the cow. In theory the idea is sound; in practice some bulls refuse to serve from a ramp, particularly if they have never done so when young.

The site of the bull pen should be carefully considered.

Bulls can become frustrated and lonely. In such circumstances the habit of masturbation may develop which may lead to infertility. If bulls can take an interest in their surroundings the chances of infertility may be reduced. For this reason it is an advantage to have tubular metal rails for the bull yard and to site the building so that the bull can see what goes on around him.

The bull box itself should be built solidly and provide a sheltered retreat from the rigours of winter and the hot sun of summer.

As a calf, a bull should be taught to lead (that is be trained to accept a halter) and to recognise man as his master and friend. At ten months of age he should be rung.

This is done by punching a hole in the septum of the nose large enough to allow a copper ring of 51–63 mm diameter to be inserted. Self-piercing rings can be used if a suitable punch is not available.

When being led as a youngster, a short piece of rope through the ring enables the bull's head to be held up; with older bulls a bull staff is used for the same purpose. A bull charges with his head down, and he can do little harm to the stockman provided his head is held high.

Handling is Essential

Regular handling of bulls is essential; if this is not done as a routine they may become resentful and truculent. I remember one breeder who refused to have automatic water bowls fitted to his bull pens simply to ensure that the bulls had to be led to water daily and thus had regular exercise and regular handling.

It is most unwise to take liberties with older bulls as they often suffer sudden deterioration of temper, and once a bull knows his strength and appreciates that the stockmen are afraid of him, he is potentially dangerous. Such a bull should be fitted with a mask to obscure his vision to ensure the safety of the men who handle him.

Preventing Footrot and Other Health Problems

In addition to proper handling, a bull requires certain regular attention to health details and, of course, proper feeding. These matters are discussed below.

Unless a bull is adequately exercised on a hard surface, his feet will grow out of shape. This tends to make service painful, particularly if the hind feet are affected. Hooves should not be allowed to grow long and proper trimming of the feet should, where necessary, be a regular routine.

When a bull is regularly exercised it will pay to walk him through the cow's footbath as a preventative against footrot. Where regular exercise is not possible, it is a good plan to have a footbath built beneath the foot-trimming stocks.

Incidentally, a bull should be introduced gradually to these devices in the same way that a heifer is introduced to the milking shed. Begin by fastening the bull up in the pen a few times so that he becomes used to it before the regular operation of foot-trimming starts.

In the winter, a bull often becomes infected with lice. I recommend that every autumn, the neck, withers and spine of the bull should be clipped to reduce the risk of attack, and that monthly dusting with insecticide should continue all winter, paying particular attention to the top of the head, behind the ears and the root of the tail.

In summer a proprietary fly spray will help to reduce worry from flies and possible bad temper, particularly where bulls are tethered in the open without shade.

Tethering is commendable as it affords the bull both freedom and change of diet. It is simple enough in practice, whether you use a specially designed tether or merely a steel rope anchored to a heavy concrete block. The tethering chain should not be too long otherwise it will foul the bull's forelegs.

The usual practice, during the grazing season, is to tether a bull for four to six hours in the daytime and return him to his pen at night. This eliminates the need for providing water while the bull is tethered.

In hot weather it is a good practice to tether by night and return the bull to his box during the day. This prevents heat exhaustion and considerably reduces fly worrying.

FEEDING GUIDE FOR BULLS

Bulls are much heavier than the average cow of the breed. A typical daily ration for a bull in mature breeding condition would be:

0.75 kg good hay } per 50 kg liveweight
2.0 kg silage or kale }
plus 2–3 kg concentrates.

Thus a bull of 600 kg liveweight would receive daily:

9 kg good hay,
24 kg silage, kale or cabbages etc.,
and 2–3 kg concentrates.

The concentrates can be the same as that fed to the dairy herd or a mixture, with a level of 16–18 per cent DCP, made up as follows:

Oats and/or beet pulp	2 parts
Barley and/or flaked maize	1 part
High-protein cake	1½ parts
or White fishmeal	1 part.

A bull should not be fed immediately before use, as he works better on an empty stomach. Neither should he be fed heavily on low-quality hay or straw, otherwise he will tend to become too paunchy and slow in service.

During the grazing season when a bull is tethered on good-quality grass, he should still receive a little hay—up to, say, 3.2 kg daily—and about 1.4 kg concentrates.

The feeding standards generally available apply only to dairy and beef cattle. The rule-of-thumb feed guide given here should be sufficient to keep a bull fit and lean for breeding, not fat. The suggested intake of average hay or silage should maintain this condition. Concentrates will be required when in work, but are not necessary at times of being laid off.

A valuable bull deserves good treatment and must get it if he is to have a long useful life. Much of the infertility in bulls is due to faulty nutrition.

Age for Service

A young bull may be used sparingly from ten months until he is sexually mature—at about two years of age. In the first year, most young bulls can be safely mated with, say, twenty bulling heifers giving the possibility of early progeny testing.

Over-use of a young bull may lead to temporary impotence or even complete sterility. Four matings per week would be the limit.

If sufficiently well fed, older bulls can be worked harder and in many herds which aim at autumn calvings, the period of greatest activity for the bulls is in mid-winter. After a long period of inactivity a bull may be temporary infertile for a few services owing to low-virility semen. After a rest period it is probably wise, therefore, to allow two services per cow rather than just one for the first few days.

Timing of Service

A cow is most likely to conceive from service towards the end of the heat period: this is explained by the fact that the egg is released from the ovary and has had time to move down the fallopian tubes to meet the sperm.

ARTIFICIAL INSEMINATION

The Milk Marketing Boards operate 9 centres and 64 sub-centres in England and Wales, 1 centre and 7 sub-centres in Scotland, and 1 centre and 3 sub-centres in Aberdeen. The Aberdeen and Scottish centres also serve the North of Scotland area. There are also an increasing number of commercial AI centres.

In recent years there has been a development in the DIY (Do-it-Yourself AI). Farmers wishing to do DIY must attend an approved training course on the subject and obtain a license before they can carry out inseminations.

DIY inseminators may, however, only inseminate cows on farms to which their license refers. DIY operators will have their own storage facilities on the farm with semen being supplied by their local AI centre.

The increase in DIY and the decline in the number of dairy cows has resulted in the decline of the official AI service over recent years. The long established procedures of the AI service still operate, however. Notification for a service should be given to the relevant AI centre before 10am, and the insemination will usually be carried out that day.

Advantages of AI

Since 1945 the development of AI has been extremely rapid in the United Kingdom. The major breed demand at the AI centres run by the Boards in England and Wales was given in Table 1.6 (page 15).

Over the past thirty years, the most significant feature has been the growth in demand for Friesian/Holsteins at the expense of the other dairy breeds.

The advantages of artificial insemination are greatest in small herds where the services of an expensive bull cannot be fully utilised. Equally, on small heavily-stocked farms the use of AI frees accommodation and food which can be used to keep another cow, thus raising income without additional capital expenditure.

The Board has for many years provided a service for the collection and freezing of semen from privately owned bulls either at one of their centres or on the owner's farm. This service is of great value as an insurance against accidents or death with a particularly valuable bull.

Test Mating

Ideally the semen used for AI service should be from sires that have given a progeny test of outstanding merit. Such proven sires are hard to find, and still harder to purchase from privately owned herds. However, the Board is prepared to buy or hire such bulls provided that, in addition to being fully progeny tested, they also pass a veterinary examination for health and fertility, including the TB and Agglutination (Abortion) Tests and semen examination.

The advisory committee attached to each centre and the Livestock Husbandry Advisers of the Ministry of Agriculture also inspect the bull and his progeny or near female relatives.

Because of the shortage of proven sires, young bulls have to be employed by the AI service in order to meet the rapidly expanding demand. The Progeny Testing Scheme is used to comprehensively test new bulls.

Breeding alone will not raise yields in a low-producing herd if management is at fault. In any case, it is a matter of some years before any herd is completely made up of genetically superior cattle. Better breeding must go hand in hand with better standards of nutrition, hygiene and general management.

Future Developments

Every encouragement must be given to the maintenance of the AI stud at the very highest level of performance. This depends on the quality of the bulls being reared. The Board is rearing bulls at Chippenham from selected matings of superior AI bulls on proven dams in pedigree milk-recorded herds, but the majority of bulls are still privately reared.

To encourage the private breeder, a system of royalties payable to the breeder has been suggested after a bull has been proven in the AI service.

Recent developments have been the formation by groups of prominent British Friesian breeders to privately test bulls of their own breeding, with the co-operation of the MMB. This is a further means of ensuring that the livelihood of the bull breeder is not jeopardised by the increasing use of AI.

Milk recording must be encouraged if the testing of young bulls is to be expeditiously carried out.

Imports of cattle from overseas include Charolais, Canadian and American Holstein, Guernsey and Jersey bulls. In addition, a number of other European beef breeds, such as the Simmental and Limousin, have been introduced over the past few years.

All of these imports are prompted by the aim to improve the cattle in this and other countries by making use of AI and progeny-testing techniques.

CHAPTER 15

DISEASE IN THE DAIRY HERD

A national committee reported many years ago that 12 per cent of total annual production was lost every year by British dairy farmers through disease in the dairy herds. Nowadays tuberculosis has been eliminated and the campaign to eliminate brucellosis is almost completed. Nonetheless, dairy farmers still have problems, including lameness, infertility and mastitis.

The reduction of losses attributable to disease and the promotion of positive welfare, health and longevity in dairy stock should be a cardinal principle in dairy herd management. Disease makes for uneconomic production and renders any breeding programme incapable of realisation.

FOUR MANAGEMENT RULES

It is not possible to give in one chapter of a book all that a dairy farmer should know about cattle diseases and their control and treatment. Readers are referred to the appropriate textbooks for such details. Farming Press produce some very good books for stockmen: *Cattle Ailments, Calving the Cow and Care of the Calf* by E. Straiton, and *A Veterinary Book for Dairy Farmers* by Roger Blowey. Nevertheless, I do want to suggest here how enlightened herd management can prevent disease.

A good stockman should:

- Be capable of detecting the early symptoms of ill-health in dairy cattle.
- Take steps to eliminate the sources of disease on the farm—whether from contaminated water supplies, infected buildings or pastures.
- Enlist the aid of the veterinary profession to control disease by suitable preventive measures.

- Adopt all management practices which will reduce the risk of disease.

Signs of Ill-Health

A mental picture of the cow in good health is a valuable attribute of good stockmanship, and the following guide is suggested to help in the detection of ill-health and prompt treatment.

The general posture of the cow, her movements, breathing and behaviour should be normal. Warning symptoms which the cow may exhibit include standing with her head down, showing undue lassitude or a tendency to separate from the herd.

Her appetite should be normal. If it is not, then make sure the cause is not unsuitable food, dirty feeding troughs or lack of water.

The skin, which should be handled over the last rib, should feel soft and pliable. A tight hide and a lack of bloom often accompanies digestive troubles or feverish conditions.

The udder should be soft and pliable to the touch rather than hard and swollen. The fore milk should be examined to see whether it is watery looking.

Nostrils should be moist and free of mucus; the eyes should be full, not sunken, and free of a fixed or staring look which often accompanies the onset of milk fever.

The dung should be reasonably formed, but will obviously be affected by diet; it should be free from gas bubbles as in Johne's disease or liverfluke.

Constipation or scouring should be particularly noted. The vulva and tail should show no evidence of coloured discharge from the genital organs; a clear discharge is a sign of the normal reproductive cycle; pus-containing discharges indicate a septic condition.

Urine should be straw-coloured. If it is darker, lighter or contains blood, urinary tract infection should be suspected.

Any fall in milk yield indicates that the animal is suffering from an interference in her metabolic processes such as wire in the rumen, displaced abomasum or any disease accompanied by a high temperature. Abnormal smells are indicative of acetonaemia or aromatic feeding agents, such as kale or turnips.

When a cow seems off-colour, take her temperature by inserting a clinical thermometer into the rectum. Normal temperature is 38.8°–39°C. The pulse rate can be taken by pressing on the artery at the base of the tail with the index finger of the right hand. Normal rate is 60 per minute.

Diagnosis

The full diagnosis of disease is, of course, the veterinary surgeon's job, but Table 15.1, which lists the symptoms of various diseases, will form a useful guide for stockmen. The body structure of the cow is shown in Figure 15.1. This will help to locate the situation of the various cow ailments.

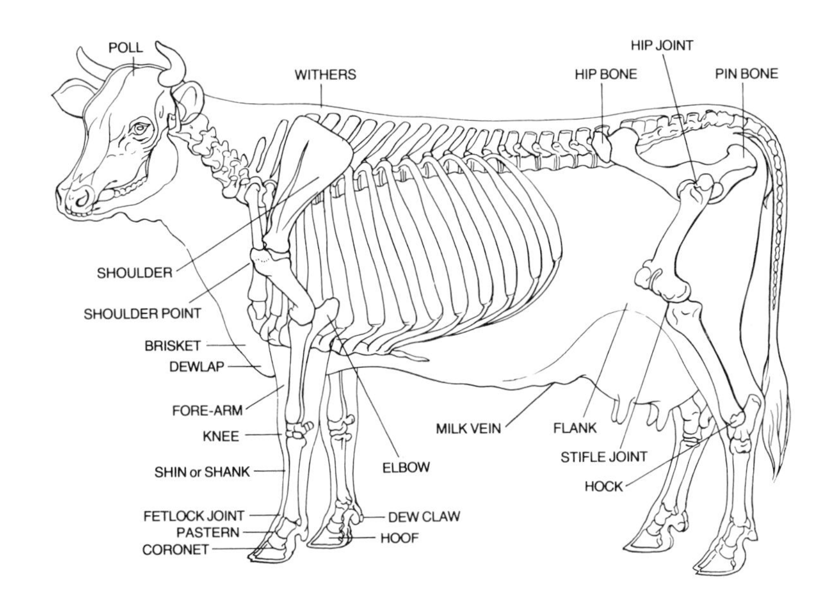

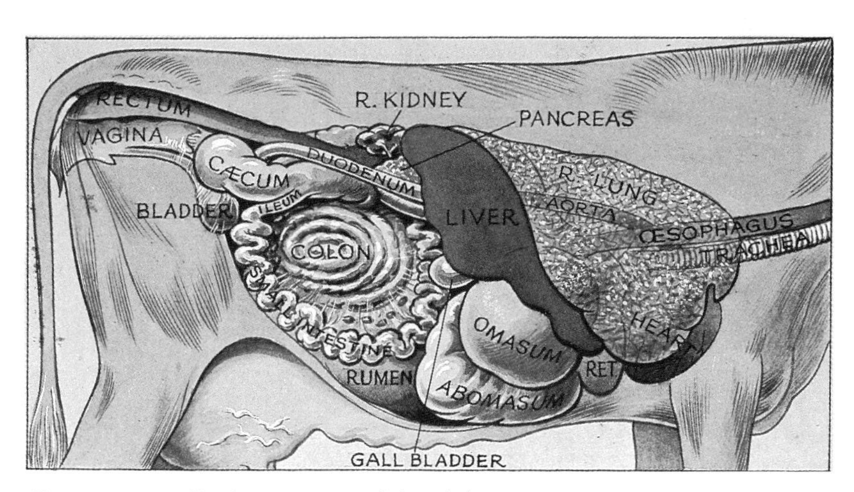

Figure 15.1 Body structure of the dairy cow

Table 15.1 Ill Health, Symptoms and Related Conditions

Symptom	Associate with:
Loss of appetite	In calves: (a) scours (b) virus pneumonia (c) joint ill (d) any other infection causing a high temperature In cows: (a) mastitis (b) wire (c) pneumonia (d) displaced abomasum (e) any condition causing a high temperature
Depraved appetite	Suspect mineral deficiency or lack of some dietary constituent
High temperature	(a) pneumonia (b) acute mastitis (c) peritonitis (d) metritis
Low temperature	(a) terminal condition following any of above (b) milk fever (c) toxic mastitis
Unsteady gait	(a) milk fever (b) staggers (c) tetanus (d) BSE
Skin tight, lacking bloom	Any condition causing scouring (a) worm infestation (b) liver fluke (c) corn poisoning (d) Johne's disease
Constipation, faeces hard and solid	(a) improper feeding, excess fibre (b) acetonaemia (c) many feverish conditions
Diarrhoea	(a) worm infestation, liver fluke (b) improper feeding, minimal fibre

Table 15.1 (contd.)

Symptom	Associate with:
	(c) Johne's disease (d) corn poisoning
Milk abnormal clots or pus in fore-milk, udder induration may be present	Chronic mastitis
Milk yield greatly reduced udder hot and painful, pus in milk, temperature high, udder induration present	Acute mastitis
Discharges from vulva	
(i) clear slime	Indicative of normal reproduction cycle
(ii) pus present and foetid smell, fertility impaired	Retained afterbirth if recently calved or abortion has occurred
(iii) white discharge, fertility likely impaired	Chronic inflammation of the uterus or vagina Infectious bovine rhinotracheitis (IBR) Campylobacter (= Vibrio) infection
Lameness	(a) foul in the foot—infection between claws (b) foot rot (c) traumatic penetration of horn (d) hereditary factors producing soft horn or malformed feet (e) concrete—newly laid (f) joint or navel ill in calves (g) Laminitis (h) Digital dermatitis (i) Sole ulcer
If accompanied by profuse salivation	Suspect (a) foot and mouth disease (b) IBR
Abnormal heat periods	(a) cystic ovaries, leads to persistent bulling

Table 15.1 (contd.)

Symptom	Associate with:
	(b) persistent corpus luteum no heat observed (c) infection following cleansing (d) IBR
Abnormal smell to breath or milk	Suspect acetonaemia if a sickly and penetrating smell
Abnormal smell plus loss of appetite	(a) wire (b) displaced abomasum
Laboured breathing	(a) suspect pneumonia (b) IBR
Loss of cud, profuse salivation	Examine for abnormal teeth; wooden tongue or jumpy jaw; foot and mouth disease
Bloodstained urine	Redwater or kidney inflammation
Blood in milk	Suspect burst blood vessel
Failure to come on heat	(a) seasonal—late winter (b) nutritional—energy deficit (c) persistent corpus luteum (d) metritis (e) poor observation (f) IBR
Unable to rise	(a) milk fever/hypomagnesaemia (b) injury to spine + nerves during calving (c) damage to pelvis during physical riding (d) BSE
Swollen joints	(a) navel ill, joint ill (b) foul in the foot, physical damage (c) mastitis
Persistent coughing, loss of condition	Suspect worm or husk infestation, house immediately, feed generously, dose on veterinary advice. Could be chronic pneumonia

Seek Veterinary Advice

Any obviously sick animal should be put in a well-littered loose-box as soon as observed. If the temperature is abnormal, keep the patient warm—if necessary by rugging up—and feed on a laxative and nutritious diet: for example, bran mashes with a little linseed cake to increase palatability.

If there is any doubt as to what ails the animal, call in veterinary advice and act upon it.

Clean Up Disease Sources

The importance of eradicating sources of infection has been stressed. Water supplies, if allowed to become contaminated with urine or faeces, are a potent source of disease. Installing a piped water supply and fencing off all open pools is well worthwhile.

Buildings should be conscientiously spring-cleaned as an annual routine. Walls and floors should be rendered with a smooth surface cement finish and woodwork replaced, where practicable, with tubular steel fittings. All cement facing and floors and all metal fittings should be scrubbed thoroughly during the spring-clean with washing soda added to the water at the rate of 1 kg washing soda in 100 kg water. Any woodwork is best either creosoted or disinfected with a blow-lamp. Sunlight is the enemy of most disease organisms—let it in wherever you can.

Pastures should not be consistently over-grazed with cattle. If heavy stocking is obtained by strip grazing, then alternate the grazing with cutting to reduce the risk of parasitic infection.

Remember, too, the advantages of mixed grazing in this respect.

Dosing Against Worms

Dosing of young cattle with an anthelmintic is recommended in early July and ten days after housing in the autumn (allowing immature worms to develop and be killed). Between these two periods, it is necessary to dose as stocking rate and condition of pasture dictate. Parasitic infection of cattle is more common than is generally realised. Again it leads to a low resistance and allows other diseases to get a hold on the stock. Husk (parasitic bronchitis) can be treated in the same way as worms—by controlling the animals' environmental conditions. Prevention is better than cure; an oral vaccine is recommended before turning out to grass.

THE COW'S REPRODUCTIVE CYCLE AND ASSOCIATED PROBLEMS

The control of disease in the dairy cow obviously plays an important role in achieving a satisfactory level of quality milk. As levels of production are increasing so the particular importance of regular breeding in the dairy cow has become recognised.

It is essential that a cow should produce a calf each year, that is, that it should have a calving interval of something approaching 365 days. To achieve this kind of breeding regularity it is fundamental that every dairyman must understand the principles of the cow's reproductive cycle. The layout of the reproductive system is shown in Figure 15.2.

An outline of the basic cycle is as follows:

1. The cow has two ovaries which release an egg on a 21-cycle. The egg development and release is connected with the presence of the follicle stimulating hormone (FSH) produced in the pituitary gland.

This follicle produces oestrogen, present in the egg, which is responsible for the signs of heat shown by the cow; it also prepares the uterus for the eventual reception of the fertilised egg. The egg is fertilised on its way to the uterus from the ovary, by way of the connecting tube, the oviduct.

2. The departure of the egg from the ovary leaves a cavity which is filled by the rapidly growing Yellow Body or Corpus Luteum.
3. As it grows, the corpus luteum secretes the hormone Progesterone which has three important functions:

- Its presence in the cow's bloodstream is necessary for the maintenance of pregnancy.
- A high level of progesterone prevents the cow coming on heat, i.e. prevents oestrus. This is the reason why cows which are pregnant do not come on heat.
- If the released egg is not fertilised it will then pass out of the cow's reproductive tract. The reducing level of oestrogen causes the corpus luteum to contract, resulting in a fall in the level of progesterone and the start of the next period of heat.

This brings the reproductive cycle round full circle in 21 days.

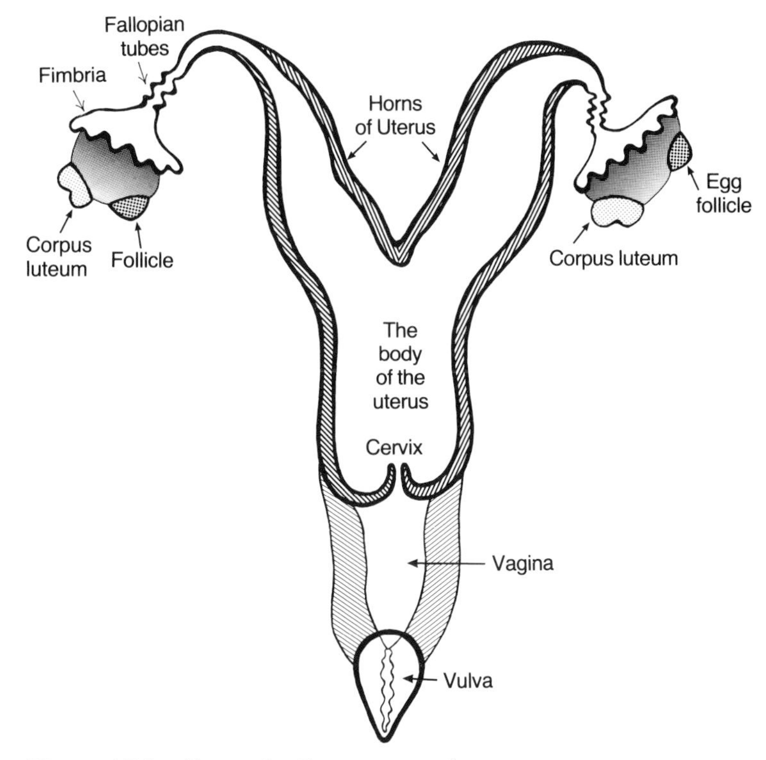

Figure 15.2 Reproductive organs of a cow

4. With the decline of the corpus luteum, the ovary begins to develop a swelling, described as a follicle, which contains the next egg to be released.

In a normally well-fed and well-managed dairy cow the reproductive cycle continues at the 21-day interval until such time as the fertilisation of the released egg is effected by service and pregnancy commences.

The successful functioning of the cow's reproductive cycle is dependent upon a highly complex balance of hormones in the bloodstream which is considerably affected by the quality of cow nutrition and management. The highest proportion of cows culled from an average herd are sent off because of their failure to breed regularly. In many cases this may not be the result of the cow's genetic ability to breed, but is due to a failure in the nutrition and a general lack of understanding or management of the cow's reproductive cycle.

Monitoring Low Fertility

This should be a basic operation of dairy husbandry and one area where theoretical knowledge must be linked to practical application. Low fertility must be recorded on a day-to-day basis; each cow acts independently from all other cows in the herd and so information on its particular breeding cycle must be recorded and acted upon. Breeding records must be kept. On many farms this takes the form of a Breeding Board, of which there are several types, both round and square. The following basic details need to be recorded, assuming of course that a 365-day calving interval is the desired objective.

1. After calving, all cows should be regularly observed and the dates when they were on heat recorded in their individual records. This must be done even if there is no intention of serving a cow, since it is important to establish the fact that the cow is cycling regularly.
2. If a cow is expected to calve once per year (every 365 days), given a pregnancy period of approximately 280 days the animal must be put in calf, i.e. served, within the first 85 days (three months) from calving. In practical terms this means effective service at the second or third heat period.
3. After service, it is vital to note whether any cow returns on heat, indicating that the previous service was not successful. Individual cows returning on a regular basis would warrant veterinary examination.
4. If cows are not seen on heat after service it cannot be assumed that they are in-calf. In the interest of efficient dairy management it is advisable to have all served cows pregnancy-tested by a veterinary surgeon. This would usually be carried out about six to eight weeks after service. Some vets with large dairy practices become very skilled at this operation and can successfully confirm pregnancies as early as five weeks.
5. Once pregnancy has been confirmed, the expected calving date and drying-off date can be established; in this way, breeding performance can be monitored during early lactation.

To emphasise the importance of this subject of herd fertility, it is worth noting how a good herd with a high level of 7,000 litre average and a calving interval of 420 days, only looks like an average herd when the yield is corrected to a 365-day calving interval.

i.e. $\frac{7,000}{420} \times 365 = 6,083$ litres

Calf numbers per year are also considerably reduced if the interval exceeds 365 days.

When cows fail to breed, the possible causes must be investigated as soon as possible.

THE IMPORTANCE OF A LONG PRODUCTION LIFE

A long milking life for dairy cows should be as important an objective as high yield per lactation. It is a disturbing fact that only about 8 per cent of recorded cows survive until the fifth to seventh lactation when production is at its maximum.

Various surveys of the extent of the replacement of dairy cows in milk-recorded herds have indicated an annual turnover of from 20 to 25 per cent. This means that, on average, a cow is in the herd for only four or five years. Yet herds are known where average working life of cows is seven or more years.

This relatively high turnover rate is due, to a considerable extent, to losses from diseases mentioned in Table 15.1 which account for at least 50 per cent of the total disposals. Losses through accidents and through deaths not directly attributable to disease are no more than 20 per cent, and disposals on account of poor milking ability amount to roughly the same.

To reduce herd wastages we therefore need:

- A better level of management to reduce the incidence of diseases and a greater disease consciousness on the part of stockmen, particularly in regard to infertility problems.
- Better breeding policies so that the general productivity of home-reared replacements reaches a higher level.
- Greater attention to the welfare of the individual cow to reduce the risk of accidents and to maintain the cow in a fit condition over a long and useful working life.

In this last respect, several points are worth emphasising. Many cows break down as milk producers through the stockman's failure to attend to their feet; hoof trimming should be a regular routine in dairy herds whenever necessary.

VALUE OF A FOOTBATH

Where lameness in cows is a recurring problem, it is probable that germs causing necrosis of the foot tissues are being picked up in muddy gateways or at drinking trough surrounds.

To combat this trouble a shallow footbath should be constructed through which the cows can be driven. If the bath is filled with a solution of 5 per cent formalin and the cows are walked through it once a week, the trouble should be cleared up.

The bath should be kept covered when not in use; otherwise rainwater will dilute the disinfectant.

Footbath immediately outside the parlour exit; one without a lip might be preferred.

BSE—BOVINE SPONGIFORM ENCEPHALOPATHY (MAD COW DISEASE)

BSE is a disease of the nervous system which was first diagnosed in the UK in 1986, and is now spread quite extensively through the south of England. It is proving very difficult to find the causal agent and means of spread of the disease. It is thought however that there might be some connection with the scrapie condition in sheep.

The symptoms include: variable manifestations of the brain

disorders especially incoordination leading to recumbency, temperament changes, peculiar head carriage, kicking, ear drooping, weight loss, licking, fear and high stepping.

Confirmation of the disease depends on post mortem histology on the brain: there is no test for live animals. There is no treatment to date.

The disease can be confused with hypomagnesaemia, nerve acetonaemia, nerve damage or listeria. Legislation has been passed to prevent the feeding of ruminant bio-products (i.e. offal) to ruminants. Evidence suggests that in every case the animal has eaten the infective agent during its own young life.

It is possible that vertical transmission (i.e. cow to offspring) may occur. This has yet to be confirmed.

Legislation was passed in 1988 making the disease notifiable and authorising a compulsory slaughter policy of infected animals with compensation to the owners. In the twelve months ending June 1989, 4,500 cases were confirmed and 7,000 cattle found infected on 4,000 farms. The disease is predicted to peak by the mid-1990s and to be virtually eradicated by the turn of the century. The disease is having serious financial consequences on cull cow and calf sales.

IBR: INFECTIOUS BOVINE RHINOTRACHEITIS (COW FLU OR RED NOSE)

This is a highly infectious disease which is becoming increasingly prevalent. It is caused by a virus and can result in:

(a) Respiratory disease
(b) Abortion
(c) Encephalitis in young calves
(d) Vulvovaginitis.

All ages and breeds are susceptible but it is mainly seen in cattle over 6 months of age and is probably more common in beef animals and during the winter months. This is due to the fact that the disease is spread by droplets, making transmission more likely while the cattle are housed. The symptoms are fever, conjunctivitis with lacrimation, and salivation. Sometimes an accompanying pneumonia may be present which is usually associated with a secondary bacterial infection.

The disease itself can be severe and the resulting loss of milk and flesh, losses through abortion and the high cost of treatment, make it a disease of increasing economic importance. Recently an intra-nasal vaccine has been produced which is highly effective even in the face of an outbreak. Duration of the vaccine is in doubt; animals should be vaccinated annually just before housing.

LEPTOSPIROSIS

This disease is a bacterial disease of cattle caused by *Leptospira hardjo*. It has become the major diagnosed cause of abortion in cattle. It can also cause infertility, weak or stillborn calves, reduced milk yield and 'flabby bag' syndrome. It is infectious to humans, and causes symptoms similar to flu although it can develop into fatal meningitis. The disease was declared notifiable in 1980.

Diagnosis is by examination of abortion material samples and blood tests. It should be noted that chronically infected carriers, which are very difficult to diagnose, exist. There is a response to antibiotics but control is usually effected by a vaccination programme. Vaccinations should be boosted annually.

Herds wholly free of *L. hardjo* can be registered as an Elite herd through the Cattle Health Scheme of the MAFF.

It is a disease which tends to be of particular interest for pedigree breeding and the export market.

GLOSSARY OF TERMS RELATED TO VETERINARY PRODUCTS

Modern veterinary science has inevitably created a whole new vocabulary of terms—the following glossary has been included so that the stockman may have some knowledge and appreciation of the terms found in everyday use.

Antiserum

This is derived from the blood of animals which have been exposed to a disease organism. The animal's immune system responds to the presence of the organism to produce antibodies, this process is called **active immunity**. These antibodies, if taken from one animal and placed into another, provide a short term of immediate immunity. This is called **passive immunity**, and is similar to the immunity provided to a calf by colostrum.

Antisera are used in the presence of a disease to provide resistance to that disease for a short period only.

Vaccines

These are used to stimulate the immune system of the body, that is, active immunity. This is achieved by giving the animal a dose of the organism producing the disease, but the organism has been modified or even killed in order that the immune response is obtained without the disease occurring. These have to be used before an animal is exposed to a particular disease risk. An example is lungworm vaccine.

Antibiotics

These are drugs used specifically to attack and kill bacteria. They may be broad spectrum (attack many different bacteria) such as tetracyclines; or narrow spectrum (attack only certain types of bacteria), for example, penicillin.

Sensitivity Tests

These are tests where a particular bacteria obtained from the infected animal is grown under laboratory conditions in the presence of certain antibiotics—this is in-vitro testing. Where no growth of bacteria occurs then the organism is said to be sensitive to that particular antibiotic.

Parasitic Control

Is in two forms:

Ectoparasitic. This is used to control external parasites, e.g. lice, mange mites, ticks.

Endoparasitic. This is used to control internal parasites. The most common term for drugs used here are 'wormers' or anthelmintics. Such wormers have varying degrees of activity against internal parasites and advice should be sought before using a particular wormer.

Methods of Drug Administration

There are three principal methods:

1. Topical: applied externally to the body surface, e.g. dusting powder, udder cream.
2. Oral: applied by mouth only, e.g. various wormers, lungworm vaccine.
3. Parenteral: applied by injection only, this may be
    - Intramuscular—into muscle
    - Subcutaneous—under the skin
    - Intravenous—into a vein
    - Intraperitoneal—into the abdominal cavity.

CHAPTER 16

INTRODUCTION TO DAIRY MANAGEMENT

Dairy farming has seen many stages of development since it started. Originally, all dairy products were sold in the local market towns to the local inhabitants. The Industrial Revolution in the eighteenth century resulted in a massive expansion of population in the industrial centres. Towns and cities rapidly grew to sizes which could no longer be adequately catered for by the local agricultural production. With the increasing demand for food, outlying agricultural areas were called upon to supply these additional food supplies. The milk production industry began with town dairies but the demand quickly outstripped supply thus furthering the spread of dairying far beyond existing boundaries. Modern transport provided a vital link to this expansion story.

By the late 1920s, however, the situation had developed to a point where the supply began to exceed demand. Prices became less stable and thus the Milk Marketing Board was set up in 1933 to cope with this type of problem. The Government of the day also set up Agricultural Economic Survey Departments at various Universities throughout England and Wales. The work of these departments was to investigate the costs of milk. This information was then utilised over the years as a basis for negotiations in determining the price of milk. During the Second World War (1939–45), imports of food from overseas were severely restricted. The policy adopted at that time was to utilise food from our own resources at any price.

In more recent years the costs incurred in milk production have increased with the level of inflation and today's situation is one where dairy farmers have to be far more aware of the business aspect of milk production, as well as the basic requirements of dairy cow husbandry. In past years, emphasis on dairy farm management was related to which cows to breed and feed, together with the growing of appropriate crops on the farm. The term now has to apply additionally to the financial aspects of running a farm.

THE LANGUAGE OF DAIRY FARM MANAGEMENT

The interest in dairy farm management started with the work of the Agricultural Economics Department. Previously, dairy production had been measured in terms of physical output and input, that is, litres of milk and kg of concentrate. Milk Recording officially started to be recognised in the early 1930s, the importance of recording feed input has not yet been widely appreciated.

The new subject of dairy farm management required a whole new set of yardsticks. Measures of economic efficiency were designed and any ambitious person working in the dairy industry today needs to understand the new, rapidly developing technical vocabulary related to this subject. As this book is concerned with the principles of dairy farming, so it is necessary for the idea of the total dairy farm business to be introduced in this chapter. The matter is dealt with comprehensively in *Dairy Farm Business Management* by Slater and Throup, published by Farming Press. However, good dairy husbandry, to be effective and profitable, must be associated with some understanding of the economics of dairy farming. As a first step, it is important to appreciate the financial structure of a dairy farm.

The Gross Margin System

The first steps in modern financial dairy management were taken during the mid-1950s by the introduction of the Gross Margin System.

The system was designed to show the relative importance of the different enterprises on a dairy farm.

Table 16.1 gives an example of a typical farm situation.

The terms used in the table are defined for a dairy farm as follows:

Gross Margin—The value of enterprise gross output less enterprise costs.

Gross Output—The value of sales of milk and calves, less herd depreciation.

Table 16.1 Gross Margin (£/cow)

Year Ending Spring	*1988*	*1989*	*%*
Milk sales	880	953	92.4
Plus Calf sales	97	114	7.6
Less Herd depreciation	42	35	
Gross output	935	1032	100.0
Less			
Concentrates	189	196	52.4
Purchased bulk feeds	11	10	2.7
Forage	78	83	22.2
Sundries	68	85	22.7
Variable costs	346	374	100.0
Gross margin per cow	589	658	
Gross margin per hectare	1143	1277	
Margin over concentrates	691	757	
Margin over feed	680	747	

Source: MMB, FMS Report No. 67, 1988-89.

Note: This report is based on a group of 112 specialist farms costed through the Total Farm Business full recording scheme operated by the MMB Farm Management Services.

Herd Depreciation—The difference between the value of animals, transferred into the dairy unit (whether from the heifer rearing unit, or from outside purchases) and the value of animals transferred out (largely cull cows).

The dairy herd is assessed on a standard valuation per cow, 'the herd basis', rather than on a 'trading basis'. Any agreed increase in cow valuation is therefore a capital matter rather than a part of normal trading valuations.

Variable Costs (Enterprise costs)—Those costs which can be easily allocated to a specific enterprise, e.g. feed to the dairy herd (which would be allocated and recorded separately from the feed used in the heifer rearing enterprise). A typical list of variable costs for the dairy herd would include:

Concentrates—purchased and home grown. Home grown concentrates should be costed at the price for which they could have been sold on the open market.

Purchased Bulk Feed—Some dairy farms have a policy of buying in bulk feeds, hay, straw, brewers grains, pressed sugar beet pulp, fodder beet and silage.

This policy may be due to a variety of reasons:

(i) When there is a very intensive stocking rate, and insufficient forage has been grown to satisfy herd requirements.
(ii) Extremes of climates, e.g. droughts which reduce grass growth or very wet seasons which force stock to be housed for longer periods than expected. In northern areas, winter can commence early and finish late; winter can easily extend from 180 to 210 days. Wet springs in particular necessitate a longer housing period since poaching of wet pastures can cause serious reductions in summer grass production.
(iii) In recent years, a system of buffer feeding, i.e. feeding silage to cows at grass, has become standard practise. This is likely to use up existing fodder supplies and lead to the purchase of additional quantities.

Forage—This category of cost would cover seeds, fertilisers and sprays. The forage costs are very much associated with the stocking intensity of the farm.

More stock lead to greater quantities of grass and conserved grass being required. This requirement can only be satisfied by short-term, productive leys using more fertiliser. Lower stocking rates would involve a greater area of land, an extensive grazing system and overall lower fertiliser costs.

Sundries—This section would include all the other items used in looking after dairy cows. Although veterinary and medicine costs are thought to be expensive by dairy farmers, compared to other costs they are, in fact, minimal. AI, milk recording and herd societies are also individually small costs relative to total costs but nonetheless sundries account for approximately 22.5 per cent of total enterprise costs.

Once the detailed make-up of a gross margin account is understood, it is then possible to appreciate the priorities of managing the dairy enterprise.

A good gross margin is largely dependent upon a high value of sales coupled with a low level of enterprise costs.

The efficiency of a dairy enterprise can be investigated using the information contained in the Gross Margin Account. A later section will deal with this topic and the various ways of assessing enterprise efficiency.

It is important to realise that the gross margin is not a measure of profit on a dairy farm.

The profit is calculated by subtracting the fixed costs from the gross margin.

Fixed Costs

Fixed costs cannot easily be allocated to specific enterprises and do not vary in response to minor changes in the size of an enterprise. The fixed cost category includes wages, machinery, rent and other numerous smaller costs, including insurances, office expenses, subscriptions, etc.

Interest charges are also included as a fixed cost item. With the increase in interest charges over the years, e.g. 5–15 per cent, and the fact that farm overdrafts have become increasingly common, this cost has become very important.

On farms where it is planned to increase the size of a dairy unit from 80 to 100 cows it is probable that the existing machinery will be able to cope with the extra work involved, e.g. scraping yards out, making silage, etc., so fixed costs would not increase very much.

A change from 80–150 cows on the other hand would definitely result in a big increase in costs. A major increased cost would be the wages bill, since the position would be reached where additional labour would be required to perform all the routine tasks (a general rule is that one man is required per 80–100 cows).

The connection between the costs and income can be summarised

Gross Output – Enterprise Costs = Gross Margin
Gross Margin – Fixed Costs = Profit/Loss

Table 16.2 gives a comprehensive picture of the finances of a typical dairy farm.

Table 16.2 Farm and Dairy Gross Margins, Fixed Costs and Profit

Farm Gross Margins	*1988*	*1989*
Dairy herd (gross margin per cow ×	£	£
number of cows)	69,068	78,053
Young stock	3,904	4,360
Other livestock	1,581	2,406
Arable	2,980	2,800
Other farm income	3,789	5,357
Total gross margins plus other income	81,332	92,976
Fixed Costs		
Wages	12,835	13,650
Power and Machinery: cost	15,551	16,153
depreciation	5,689	5,725
Sundries	6,874	7,280
Property charges: cost	8,617	9,483
depreciation	2,150	2,308
Interest	9,302	10,038
Total fixed costs	61,018	64,637
Profit	20,304	28,339

Source: MMB—Information Report Unit No. 67, 1988–89.

PROFIT—THE MOST IMPORTANT FIGURE

The figure that every farmer is vitally concerned with is the profit at the end of the year. Once this figure is obtained it is a matter of deciding whether it represents a suitable reward for his labours during the previous year.

If he feels that the profit is not high enough, then he should investigate the farm figures to find out whether increased profits in the future are to come from reduced costs or more efficient and higher levels of output. The exercise really is in two parts:

1. Reducing costs

A detailed review of the fixed costs should indicate whether there is any scope for economy in this particular area. To help in deciding whether levels of costs are acceptable, it is necessary to obtain information from similar farm-types to compare their cost levels. Reports concerning this type of information are available from the various University Economics Departments and other commercial agencies, e.g. MMB.

There is a limit to the level to which costs can be reduced without affecting productivity. For example, a farmer and his wife might attempt to run a small dairy farm with, perhaps 100 cows by themselves, having thought that previously the wages bill was too high and therefore having made their only farm worker redundant. In such a case the farmer may be so busy carrying out his practical farm work, in perhaps a twelve–fourteen hour day, that he has no time available to manage the farm. As a result of this pressure of physical work, fewer records are kept, hence he is not so well informed as to what is happening with reference to output and related input. Decisions on buying and selling livestock, feed and fertilisers, etc., are all made with little time for comparing alternatives. Ultimately, the farm production suffers and despite the saving in fixed costs farm profits could go down.

2. Improving Enterprise Efficiency

The scope for improvement in this area can be best appreciated if brief reference is made to the gross margin account in Table 16.1. Since gross margin equals gross output minus enterprise (variable) costs, the highest margins are to be obtained by (i) increasing the output, (ii) reducing the enterprise costs, and in some cases (iii) balancing the resources better to improve effiency.

The introduction of quotas, however, restricted the scope for increasing production.

The first step is to identify the priorities of management. A detailed inspection of the gross margin account for 1989, shows up two particularly important features: (i) milk sales make up 92% of the enterprise Gross Output; (ii) purchased feeds make up 77.3% of total enterprise costs.

These calculations strongly support the idea that profits from dairy cows greatly depend upon how much milk the cows give and how much it costs to produce the milk. For many years the main accent in dairy production has been the recording of milk yield; in the future, this will need to be matched with equal interest in how to feed the dairy cow economically.

As a result of this philosophy, it is inevitable that the practical outcome would be the introduction of some form of Dairy Herd Recording, based on assessing the relationship between milk produced, feed consumed, number of cows in the herd and the size of the herd quota.

DAIRY HERD MANAGEMENT RECORDING

Dairy herd management is concerned with making decisions which, hopefully, make money. With the economic pressures on the dairy farms of today, there is little margin for error. The old fashioned method of scribbling a few figures on the back of a cigarette packet is no longer suited to today's conditions. The information which is necessary for making successful management decisions to achieve the highest profits in milk production is best grouped under three headings.

Herd Forecast of Annual Milk Production

It is important that the first step in herd recording should be the production of a plan which will provide the best possible balance between cow numbers and yield per cow to fit the farm quota.

In the immediate post quota years, there was a feeling of gloom in the dairy industry; hardly surprising since there had been an enforced cut-back in production by 10 per cent.

For a variety of reasons, however, some farmers produced less than quota, and, following legislation introduced in the late 1980s, it became possible for this 'spare' quota to be transferred to farmers who had the ability to produce more milk than their quota allowed.

This resulted in a change in planning terms, away from making the farm fit the quota, and towards the leasing or purchase of sufficient quota to fit the resources of the farm. Nonetheless, a herd forecast is still essential.

The information needed to prepare the herd forecast is shown in Table 16.3.

The basic information required for a forecast is:

(i) *Number of cows to dry off each month.*
Drying off dates are best calculated by arranging to give each cow at least eight weeks dry period before its expected calving date.

(ii) *Cows to be culled.*
This if of necessity an estimate. Although it is likely that cull cows will make up some 20–25 per cent of the herd, there is no means of anticipating which animals will leave the herd, particularly in the case of casualties and infertile cows.

Table 16.3 Data Required to Provide Herd Yield Forecast

1. Farm Name and Address:..MR...FARMER........................

2. For 1st day of .....APRIL.............. 19.89.. Please provide the following:
   (a) Herd Yield litres ....2465................ (Bulk Tank Dip)
   (b) No. of Cows in Milk ....124..............
   (c) No. of Cows Dry ....14.................
   (d) No. of Cows in Herd ....138...............

	A	B	C		D		E
Month	No. Cows to Dry Off	No. to be Culled	Expected No. Calvings **		No. to be Purchased		
			Heifers	Cows	Heifers	Cows	
April	11	3	3	7			
May	17	3	7	5			
June	13	3	2	11			
July	14	3	4	17			
August	8	3	7	13			
September	10	3	2	14			
October	10	3	2	8			
November	11	3	7	10			
December	4	3	6	10			
January	5	3	–	11			
February	7	3	–	4			
March	5	3	–	5			

** Cows and Heifers already on the Farm

4. Quota for 19.89-90
   Basic ....826,000.................. litres
   plus Leasing (if any) .120,000....... litres
   plus Purchases (if any .... litres
   TOTAL 946000

5. Average Herd Size.......145..................

Forecast gives 3·78% threshold

6. Level of Peak Yield Required Cows ....33........ litres
   Heifers ..22....... litres

(iii) *Expected number of calvings.*

This would seem to be an easy number to arrive at. Unfortunately cows do not always conceive to first service; the conception rate usually being about 60 per cent. The only way to obtain an accurate assessment is by the use of veterinary pregnancy diagnosis. Unfortunately, due to the expense, this is not a very popular management tool. There are, however, very strong reasons for recommending the practice.

An example of the importance of pregnancy diagnosis occurred in one herd, where only twenty cows calved out of an expected forty calvings in the month. This shortfall in the calving of 20 cows predicted to enter the herd and to peak at 30 litres would eventually result in the daily herd production being down some 600 litres per day (20 cows × 30 litres) and the monthly figure being 18,000 litres down on the forecast.

If this was repeated to any extent in other months then it would certainly result in the total milk production during the year falling well short of the forecast total.

A very practical use of the forecast is to decide whether the scheduled calvings at predetermined peak yields are likely to produce over- or under-production compared to the quota or quota + threshold level required.

Quota leasing has to be completed normally by the end of July, so the forecast information on milk production is ideally required by the end of June. This would allow quota to be leased if necessary.

It must be appreciated, however, that the total number of calvings calculated can only be a 'best estimate'. Information on calvings in February and March would only be provisional, services having only just taken place in May and June. However this initial forecast would indicate whether the total milk production was wildly adrift from the required amount by say ±50,000 litres. This information would be useful to have and could lead to extra animals being bought or surplus animals sold as necessary.

The reproductive cycle of cows (i.e. heat periods and services) in the herd must be recorded in considerable detail. When heifers are to be introduced into the herd, it is even more important to have pregnancy diagnosis carried out. Heifers are often allowed to run with a bull over a period of time, possibly 2–3 months in the summer. Natural service in this case is unlikely to be observed, hence the emphasis is on the need for veterinary inspection.

(iv) *Purchase of animals*
This policy is likely to be considered once the first forecast has been calculated. The majority of dairy farmers rear their own herd replacements. There may be occasions when the annual crop of heifers is lower than normal expectations. This would mean that there would be insufficient down-calving heifers to come into the herd two years later. As such it would be necessary to purchase either heifers or cows to make up the shortfall.

(v) *Quota level required*
The level of milk that a farmer wishes to produce is really a measure of what he considers would be in keeping with his farm resources, his ability to get milk out of cows and the amount of profit which he wishes to make to satisfy his ambition and financial commitments.

Financial circumstances change however, requiring policy changes from time to time.

A simple case would be where a farmer originally produced an annual farm profit which he was quite happy with. As his family grew up, two married sons came into the business requiring the farm to support three families instead of one. The farm profit must therefore be increased. On a dairy farm the obvious way is to produce more milk, more economically, possibly requiring the purchase or leasing of additional quota.

(vi) *Level of peak yield required*
There is a relationship between total lactation yield and the peak daily yield for a cow. A simple equation was devised by Dr W. Broster who recently retired from the Dairy Research Institute, Shinfield at Reading:

Cow peak yield × 200 (cow index) = total cow lactation yield.

The Broster equation was determined with reference to the normally expected lactation curve of the dairy cow based on NMR lactation research.

With developments in cow feeding it is now appreciated that lactation curves have changed over the years. Nowadays it is more likely that the cow index would be approximately 215 for cows, and 245 for heifers.

e.g. 6,000 litre lactation yield for a cow

$$\text{Peak required} = \frac{6{,}000}{215} = 28 \text{ litres per day}$$

5,100 litre lactation yield for a heifer

$$\text{Peak required} = \frac{5{,}100}{245} = 21 \text{ litres per day}$$

The cow index for a particular herd can be calculated by studying a group of cows which calved in the same month.

$$\text{The cow index} = \frac{\text{average total lactation yield for group}}{\text{average peak yield for group}}$$

The result will vary, however, depending upon the availability of bulk feed and climate. The type and quantity of feeding affects both peak yield and total lactation yield.

Over the past forty years the MMB AI service has been used extensively in the UK. Considering the quality of bulls used in AI service, it can be appreciated that the national dairy herd has a very good potential for high lactation yields. The attainment of high yields depends initially upon cows reaching the necessary peak yields as outlined above. The responsibility lies with the farmer to get the cows to achieve the required peak by appropriate feed management.

In the example farm (Table 16.4) the total herd production, with basic quota plus leasing, was 946,000 litres. With the programme of calvings of heifers and cows throughout the year, it was found necessary to peak monthly groups of cows at 33 litres, and monthly groups of heifers at 22 litres.

With computerisation it was possible to try a selection of peak yields to see which level will produce a total yield closest to that required.

The herd forecast (see Table 16.4) can be produced using the details from Table 16.3. This indicated a total forecasted yield of 981,810 litres.

Since the adjusted quota is 946,000 litres, this represents

$$\frac{981{,}810}{946{,}000} = 1.0378 = \text{Quota} + 3.78 \text{ per cent threshold.}$$

With a normal threshold of 4 per cent this would be a feasible management plan.

At the year end there was a difference between actual and forecasted production of

$$981{,}399 - 981{,}810 = -411 \text{ litres.}$$

Such a small discrepancy is, in fact, quite unusual.

There are several possibilities whereby the actual production

Month		Yield at start		Cows dried and culled		Heifers		Cows		Yield at end		Average yield per day		Yield in month	Cumulative yield
		Bulk Tank total	No. in milk	Dry	Cull	No.	per day	No.	per Day	Total	No. in milk	Total	No. in milk		
Apr.	Actual	2465	124			3	–	6		2580	124	2522	123	72834	72834
	Budget	2465	124	11	3	3	22	7	33	2390	120	2427	122	72818	72817.5
May	Actual	2580	124			5	–	7		2214	111	2397	115	74969	147803
	Budget	2390	120	17	3	7	22	5	33	2290	112	2340	116	72525	145343
June	Actual	2214	111			3	–	12		2349	110	2281	111	69965	217768
	Budget	2290	112	13	3	2	22	11	33	2324	109	2307	111	69197	214540
July	Actual	2349	110			7	–	17		2676	121	2512	116	76591	294359
	Budget	2324	109	14	3	4	22	17	33	2587	113	2455	111	76118	290658
Aug.	Actual	2676	121			2	–	13		2763	137	2719	126	83622	377981
	Budget	2587	113	8	3	7	22	13	33	2813	122	2700	118	83696	374354
Sept.	Actual	2763	137			2	–	9		2771	122	2767	126	85441	463422
	Budget	2813	122	10	3	2	22	14	33	2920	125	2866	124	85992	460345
Oct.	Actual	2771	122			3	–	10		2740	122	2755	123	87026	550448
	Budget	2920	125	10	3	2	22	8	33	2819	122	2870	124	88962	549308
Nov.	Actual	2740	122			9	–	11		2990	132	2865	128	84467	634915
	Budget	2819	122	11	3	7	22	10	33	2895	125	2857	124	85718	635026
Dec.	Actual	2990	132			1	1	6		2982	133	2986	132	90445	725360
	Budget	2895	125	4	3	6	22	10	33	3005	134	2950	130	91451	726477
Jan.	Actual	2982	133				–	7		2930	132	3051	134	91956	817316
	Budget	3005	134	5	3	0	22	11	33	2995	137	3000	136	93001	819479
Feb.	Actual	2930	132				–	5		2808	126	2869	129	80731	898047
	Budget	2995	137	7	3	0	22	4	33	2738	131	2867	134	80263	899742
March	Actual	2808	126				–	10		2486	128	2154	127	83352	981399
	Budget	2738	131	5	3	0	22	5	33	2557	128	2647	130	82069	981810

might start to diverge from the forecast, 3 months drought for example.

The most important task in managing the dairy herd therefore is to 'make the forecast happen'. If this is to occur, then the forecast should be translated into everyday action on the milking parlour floor at 6am in the morning, and for the rest of the day, every day.

Daily Dairy Herd Management

Making It Happen—as it should
Mention was made in Table 16.1 concerning the priority areas in the gross margin breakdown of dairy herd economics.

(i) *Milk Sales*— 92.3 per cent of gross output

(ii) *Feeding the Dairy Cow*	proportion of enterprise cost	
Concentrates	52.4 per cent	77.3 per cent
Purchased bulk feed	2.7 per cent	
Forage	22.2 per cent	

The two priorities for recording are therefore:

(1) Milk yields
(2) Purchased feed input

Milk Yield Recording

Traditionally, milk recording has been the tool of the pedigree cow breeder, the object being to provide an official record of individual dairy cow production with reference to milk yield and quality. This was, and still is, of importance for the provision of pedigree performance associated with the sales of pedigree cattle.

For the commercial dairy farmer, milk recording provides a means of controlling levels of milk production and of assessing whether the necessary cow and heifer peak yields are being obtained.

To simplify the control procedure, it is helpful to consider a herd as being made up of monthly calving groups of cows and heifers. The average milk yield can then be determined for the group. It is necessary to compare the average group yield with the desired peak yield during early lactation. If the required peak yield is not being attained then it is the responsibility of management to make changes in the feeding policy so that the target peak is achieved.

Peak yields usually occur between the sixth and tenth week of lactation, the exact time depending upon the level of feeding. If cows are inadequately fed then they soon convert liveweight into milk, milk yields drop and peaks could then occur between four to six weeks. Extreme body weight loss causes infertility. Underfeeding will prevent the necessary yields ever being achieved.

It has been known for cows which calve in February to reach their peak yield during May when a large supply of good spring grass becomes available.

Official milk recording is carried out by the National Milk Records Service of the MMB on a monthly basis. For dairy management purposes this is not often enough.

There are key times of the year, when it is particularly important to make correct management decisions. These can only be made on the basis of up-to-date information. The regularity of recording controls the availability of information for decision making. For example spring grass is suitable for high levels of milk from grass. As it matures into summer grass, so the level of energy and protein declines, and the fibre content increases. A management decision must be taken as to whether or not the grass should be supplemented with concentrates. A typical 'cause and effect' situation arises; would it be economic to spend 15p per day on feed to prevent the loss of milk worth 32p per day? One monthly group of cows could be fed as suggested and milk recorded a week later. The success of the feed change could then be assessed.

The milk price during the period July–October can increase by as much as +21.6 per cent (in July) to +30.3 per cent (in August). It is therefore vital to get the correct balance between milk yield, milk price and feed price for the best on-going margin of profit. Any delays in recording reduce the possibility of achieving the best margins.

Milk recording, once interpreted can be translated into feed levels greatly increasing the potential of greater profit. Milk recording in the winter is equally important. Silage is the most widely used bulk feed for dairy cows.

Silage can be analysed chemically but the only way to assess silage value in a steady progressive manner is to ask the cow what she thinks. The easiest way for the cow to communicate her feelings is by either increasing or decreasing milk yield: a continuing decline in yield suggests inadequate feeding leading to body weight loss. If this happens, it is likely that cows will also begin to have infertility problems.

A cow is designed and bred for milk production. If the level of yield is poorer than anticipated, it is very likely that management rather than the cow herself is at fault. Milk recording, then, is not just for assessing cow yields—it is also for assessing management.

Feed Recording

The major part of the dairy cow ration is bulk feed produced on the farm—grass in the summer and silage in the winter.

Very few farms have the necessary equipment to measure the feeding of bulk feed accurately since the cows tend to be fed on an ad lib (to appetite) basis.

All dairy cow rations contain a balance of bulk feed and concentrates. This balance must be controlled to make sure that the best level of margin (milk value minus purchased feed cost) is obtained.

Concentrates are much more expensive than bulk feeds so it is very important to control the quantities used. The amount needed in any situation is as little, or as much, as is needed to give the best level of margin.

Any person who is responsible for feeding cows will have to calculate the quantity of concentrates necessary to produce the required level of yield. Once this has been calculated, then the remainder of the cow's appetite can be filled with bulk feed.

After a period of feeding a ration, perhaps a month, the stockman should ask himself 'Have I fed more or less concentrates than should have been fed?' To calculate the amount that has actually been fed, requires some form of feed record.

Table 16.5 Example of Feed Record

			Dairy Cow Concentrates (Tonnes)
A		Opening valuation—1 March	5.00
B		Deliveries—15 March	9.50
C	=A+B	Total available	14.50
D		Closing valuation—31 March	3.50
E	=C+D	Amount used during March	11.00
F		Planned amount of Concentrates fed in March	8.00
G		Actual use of Concentrates compared to planned use	+3.00

If the amount used (i.e. 11 tonnes) is compared to the planned amount required (i.e. 8 tonnes) then excess usage during the month becomes apparent. In this example, 3 tonnes more concentrates were used than had been planned. At £160 per tonne, this appears to represent a misuse of £480 worth of feed.

The month's production should also be compared to forecasted production, since the extra food may have resulted in increased production reducing this 'loss'.

Once a problem has been identified then steps can be taken to remedy whatever is causing it e.g. adjustment of incorrectly set parlour feeders.

The objective of feed recording is to make sure that purchased concentrates are fed at the correct level; 'control' is the key word.

On occasions there is a shortage of home-grown bulk feed e.g. during a dry summer when grass supplies start to fail. If the DM appetite is not satisfied then the animals become restless and milk production tends to fall. To prevent this bulk foods will need to be purchased, e.g. brewers' grains, pressed sugar beet pulp, straw, hay or silage.

Although bulk feeds are much cheaper than concentrates, it is still necessary to assess usage. This is usually done as an average usage per animal in the group being fed.

Combining Milk and Feed Recording

These two recording practices make it possible to closely monitor the efficiency of daily production from the dairy herd.

The important principle behind the collecting of milk and feed records is to examine the cause and effect relationship involved in milk production.

The on-going task of managing the dairy herd is that of checking to see whether the calculated dairy ration can actually achieve the levels of milk production required by the herd forecast. At the same time the cost of the ration must be related to milk output thus ensuring that milk production is efficient, providing the highest possible on-going margin.

Figure 16.1 shows the relationship between milk production from bulk, and the average daily production throughout the year for a December-calving group of cows on a particular farm. Five points in the lactation are considered in detail:

1. Mid-February Average group yield at peak of the lactation—33 litres. Since bulk provided M + O, concentrates had to be fed

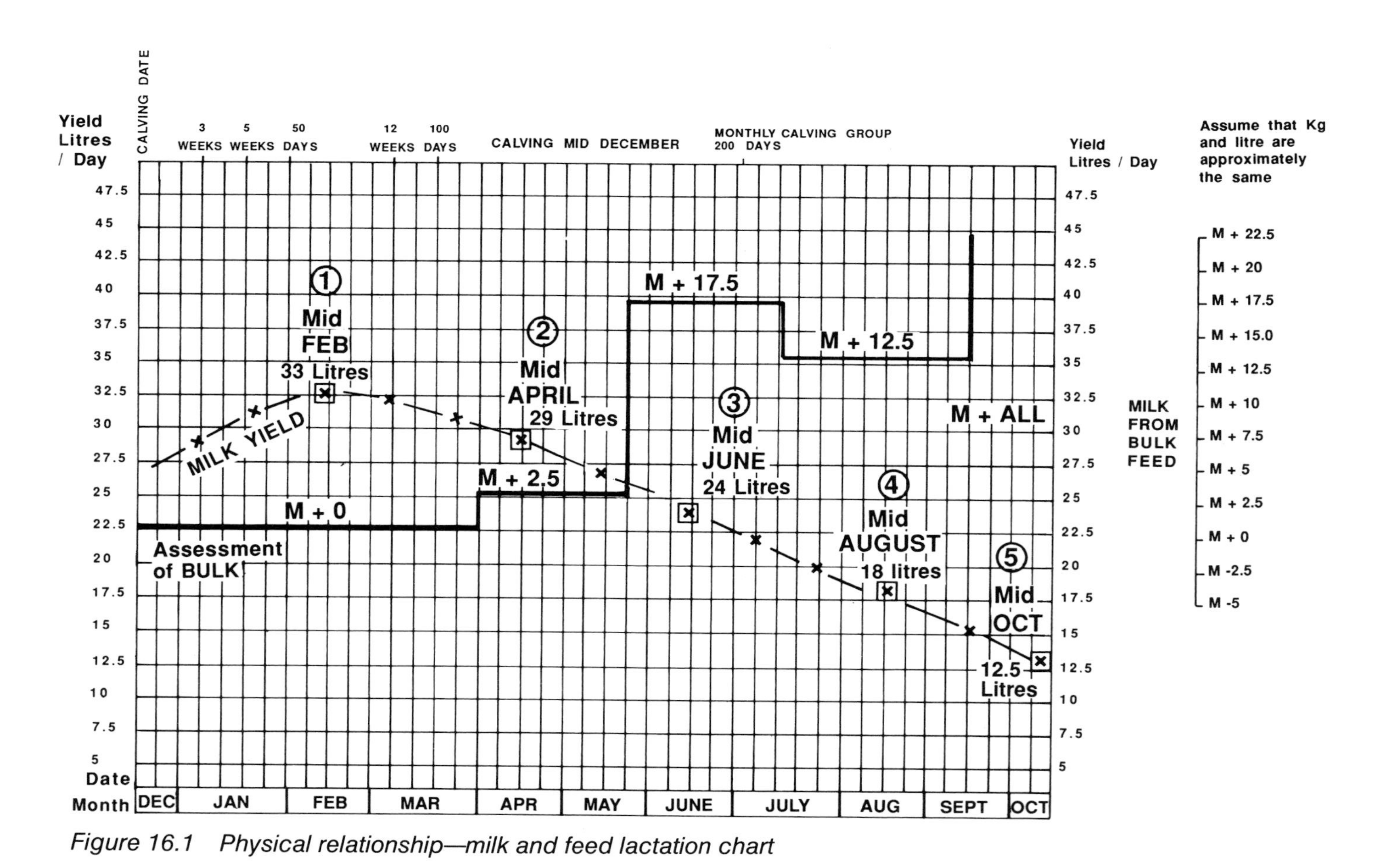

Figure 16.1 Physical relationship—milk and feed lactation chart

for every litre of production. The silage was made the previous summer under very dry conditions and therefore the quality is not as good as hoped for, resulting in the lack of milk from bulk.

2. **Mid-April** Group average yield—29 litres. The majority of cows in the group were safely in calf. The cows are giving a reasonable yield to allow good margins from grass. The seasonal price is very low but since good grass is available, the usage of quality concentrates can be considerably reduced. Care must be taken at this time to control the milk yield decline by adjusting the level of concentrate feed in balance with the value of the grass. Excessive yield decline at this time will lower potential milk sales during the high seasonal milk price period from July to October.

3. **Mid-June** Group yield average—24 litres. Lack of rain in previous weeks resulted in a poor supply of grass. In addition by early June, grass quality has gone down since the spring grass has deteriorated into summer grass. Weeks of hot weather also caused the grass to develop seed heads, resulting in increased fibre and decreased digestibility. All these factors combined to limit the amount of milk from grass compared to that available under more normal conditions. Normally, another 2½–5 litres would be produced from grass.

4. **Mid-August** Group average yield—18 litres. A late summer situation of dry weather with a few intermittent showers restricted the amount of milk from bulk feed. Nonetheless, the high seasonal milk price makes feeding an economic level of concentrates worth considering.

5. **Mid-October** Group average yield—12.5 litres. Cows being dried off through October to give the very necessary dry period of eight weeks.

At this point all production was from grass, no supplementary concentrates having been supplied since early September. A plentiful supply of grass should restore any possible loss in body condition.

From the discussion applied to the various relationships of milk and feed throughout the lactation of the December-calving group, it can be appreciated how management can use records to closely monitor progress, identifying mistakes and taking corrective action as necessary when performance starts to deteriorate.

The frequency of any recording, be it weekly, bi-weekly or just monthly, depends upon the urgency of knowing what is going on at potentially critical situations, e.g. maximum use of spring grass, feeding levels in a drought and attaining necessary peaks from newly-calved cows.

The physical relationships between milk and feed can be translated into a financial measure of production (see Table 16.6).

It is interesting to note that the margin per month does not vary as much as expected throughout the year.

From this information it is evident that a high total margin for a group of cows is only achieved by producing the best possible margin for each of the ten months when they are in milk. It is also important that the search for milk produced from bulk feed should apply to each month of the year.

Table 16.6 Milk, Feed and Margin/cow for a December-calving Group of Cows

	Months through the Lactation				
Months	*Feb.*	*April*	*June*	*Aug.*	*Oct.*
Yield (litres)	33	29	24	18	13
Milk price (p/litre)	17.57	16.65	15.25	20.11	19.43
Milk value (£)	5.79	4.82	3.66	3.61	2.42
Assessed value of Bulk Feed	M+O	M+2.5	M+17.5	M+12.5	M+15
Concentrates fed (kg)	13	10.6	2.6	5.5	—
Concentrates cost @ 14.5p/kg	1.88	1.53	0.37	0.80	—
MOC (£) /day	3.91	3.29	3.29	2.81	2.42
/month*	109	99	99	87	75

* Rounded to nearest whole number.

The average of the five monthly margins $= \frac{109+99+99+87+75}{5} = \frac{469}{5}$

$= £94$ per month

For the lactation 10 months at £94 = £940

On many farms, although much recording is carried on through the winter months, this dies down during the summer due to the increased total workload. This should not happen, in fact recording is even more important during the summer, particularly with the present high summer milk price policy.

Considerable margin can be lost through the summer. With a dry summer, the supply of grass becomes limited and the quality also deteriorates. It is quite possible that the dry matter intake of a cow could be reduced to possibly 50 per cent of what is required. The shortfall of grass must be made up by supplementing feeding with silage, a farm forage crop (e.g. stubble, turnips, or kale) or purchased brewers' grains; otherwise milk production will fall fast.

Generally it is the late lactation cows which suffer most in drought years, since they are totally dependent on grass for all of their production.

Table 16.6 provides a simple check on the margin over feed produced by a monthly group.

Taking the five months as described would show an average margin per month of £94. This extended to a 10-month lactation would suggest that these levels of margin throughout the whole lactation would result in a margin of £940 (94 × 10 months).

This exercise could be carried out on each of the monthly calving groups in the herd. An interesting comparison would then be possible to determine which was the best month to calve cows on a particular farm.

The actual rolling herd margin over concentrates on the example farm for the period April 89–March 90 was £984, which is little different to the December group total monthly margins, i.e. £940.

If all the monthly calving groups produce margins of this order, then this ensures that the total herd margin will be as expected.

DAIRY HERD COSTINGS

Dairy costings are used to measure financial and physical returns which result from the application of a high level of day-to-day dairy management moderated with reference to the requirements of the herd forecast.

Monthly Detail

This provides a regular monthly review of the current relationship between cow numbers, milk yield, feed and margin. It is useful to compare this value with the previous year's performance and thus measure progress. The monthly basic details highlight the simpler aspects of management. For example, if the milk in one month was down compared to the same month in the previous year then it would be important to find out the reason. First thoughts might be that fewer cows were in milk as a result of an extended calving pattern, alternatively, in a dry summer there might be insufficient grazing to maintain the level of milk production. Concentrate feeding on most farms tends to conform to a regular pattern. If this is disturbed, then again questions need to be asked. If higher quality bulk feed is used in summer and winter, then fewer concentrates will be used by comparison to the previous year.

Other possible reasons for altered concentrate usage are:

(i) the feeders in the parlour are not operating efficiently, particularly if they haven't been adjusted recently.
(ii) the density of the ration may have changed, affecting the amount of food in a given volume.

Efficiency factors, e.g. margin per litre, kg/litre, margin per cow and yield per cow, all offer opportunities to put the finger on mis-management.

Rolling Averages

The idea behind the rolling average is to present a 12-month production total figure at the end of every month.

On the example farm, the total herd milk yield in each month was generally higher than in the same month the previous year, resulting in the rolling average milk yield in Table 16.7 increasing from 944,905 litres in April to 981,169 litres in March the following year, an increase in yield over the year of 36,264 litres.

This was in line with the management objective, as shown previously in the herd forecast. The difference between forecast and actual milk produced was −411 litres. In order to prevent the farm going over quota extra milk production had been covered by leased milk quota.

The ability to produce more or less total milk as required is probably the largest task tackled by management in the year.

On the example farm, there were quite reasonable improvements in most of the herd efficiency factors. If all the factors were investigated in detail it would then become clear as to which areas of production still have scope for development and improvement.

Costings figures are regularly used for inter-farm comparisons. This is a useful exercise for the setting of standards, whether the average level or top-of-the-table level.

Possibly, however, too much is made of these comparisons since no two farms have identical physical characteristics.

Even in one small geographical area, wide differences can be found in: soil type (sandy-clay), rainfall (high–low), altitude (high–low), aspect (north or south).

The real value of costings therefore is in finding out what has happened on your particular farm and for giving specific recommendations as to where further improvements are possible.

Recording dairy herd details does not itself make increased profit margins, it does, however, show where improvements can be made to increase the profit potential of the dairy herd.

ASSESSING DAIRY MANAGEMENT

The various stages of costing a dairy herd have been explained and at first glance, it would appear to be a rather time-consuming task to regularly collect and process the necessary information.

Dairy farmers and workers, by tradition, work very hard for many hours each week. Unfortunately, the ability to work hard in a physical sense is not necessarily going to produce the answers for the problems of the future. The farmer/manager of the future will have to make full use of the technical information which is currently available concerning economic milk production. Future success is going to depend not only on new technology but in particular on the monitoring of any new techniques to ensure that they result in greater profitability. The explanation of a dairy costing scheme in this chapter shows the type of information which should be collected. As the information builds up year by year, so it can be studied and should give guidelines for future management decisions.

All this takes time and a priority on all dairy farms in the future must be to set aside a certain amount of 'management time' on a regular weekly/monthly basis, when the person in charge of the dairy herd can quietly sit and examine in detail the aspects of his dairy herd operation.

Many questions need to be constantly posed and answered, all relating to the information contained in the costing sheets. It

Table 16.7 Dairy Herd Monitoring Data

A. DAIRY FARM YEAR ENDING MAR'90

(a) Monthly data—Totals

Month	No. of Cows[a]			No. Culls		Calvings		Milk Sales		Feed Cost				Margin over Feed £
	In Milk	Dry	Total*			Hrs.	Cvs.	Litres	£	Concentrates tonne	Concentrates £	Other £	Total £	
April	124	18	142			3	6	72804	12126	27·08	3907	–	3907	1723 8219
May	111	27	138			5	7	74869	11280	13·50	2020	–	2020	9260
June	110	28	138			3	12	69965	10670	16·30	2473	–	2473	8197
July	121	24	145			7	17	76591	14416	19·10	3049	–	3049	3227 11367
Aug.	137	10	147			2	13	83622	16819	21·60	3445	–	3445	13374
Sept.	122	22	144			2	9	85441	16557	25·72	4010	–	4010	12546
Oct.	122	24	146			3	10	87026	16910	30·30	4637	–	4637	12273
Nov.	132	22	154			10	10	84467	15327	30·20	4568	–	4568	10759
Dec.	133	17	150			1	6	90445	16498	34·16	5195	–	5195	6605 11303
Jan.	132	18	150			–	7	91956	16314	33·66	5286	–	5286	11028
Feb.	127	20	147			–	5	80731	14448	29·10	4723	–	4723	9725
March	126	20	146			–	10	83352	17906	26·80	4305		4305	13601

[a] At end of month * Include Purchases * Average No. per year

Month	(b) Monthly per Litre				(c) Monthly per cow		(d) Twelve Months Rolling Averages						
	Sale Price p	Feed Cost p	Margin £	Kg	Margin £	Yields Litres	Total Margin £	Total No. Litres	Margin per Litre p	Margin per Cow £	Yield per Cow	Kg per Litre	No. Cows
April	16·65	5·37	13·65	0·37	57·88	512	125077	944905	13·24	864	6532	0·32	145
May	15·06	2·70	12·37	0·18	67·10	542	125915	942294	13·36	868	6499	0·32	145
June	15·25	3·53	11·72	0·23	59·40	506	126915	942106	13·47	876	6509	0·32	145
July	18·82	3·98	14·84	0·25	78·39	528	128424	942234	13·63	886	6502	0·32	145
Aug.	20·11	4·12	15·99	0·26	90·98	568	131092	950668	13·79	899	6549	0·32	145
Sept.	19·38	4·69	14·68	0·30	84·77	593	133233	957068	13·92	913	6604	0·32	145
Oct.	19·43	5·33	14·10	0·35	84·06	596	133487	957608	13·94	916	6616	0·31	145
Nov.	18·14	5·41	12·74	0·36	69·86	548	135096	957819	14·10	923	6590	0·30	146
Dec.	18·24	5·74	12·5	0·38	75·35	602	139890	964276	14·51	956	6631	0·31	145
Jan.	17·74	5·74	12·00	0·37	73·52	613	139679	971066	14·38	960	6678	0·31	145
Feb.	17·89	5·85	12·05	0·36	66·16	549	139789	976576	14·31	960	6712	0·32	146
March	18·019	5·16	16·32	0·32	93·16	570	143339	981169	14·61	984	6740	0·31	146

is important for production targets to be set. If it is possible, an attempt should be made to join a local study group. The average performance of the group should then become a reasonable objective to aim for in terms of physical production and financial return.

The next stage for any ambitious young person should be to produce the best result in the group.

Milking cows and washing parlours down does provide a certain level of physical satisfaction but it is more satisfying to know that your herd is achieving a higher and more economic yield than any other herd in your group. A standard of performance which was reached recently on one north country farm might provide a level for others to strive for:

- 7,000 litres yield at 0.26 kg/litre (2 × milking),
- stocking rate of 2.47 cows per hectare and a calving index of 374 days.

The profits of a dairy farm are dependent upon every one of several factors, some of the priorities being:

(i) milk yield, (ii) feed use, (iii) stocking rate, (iv) herd fertility and (v) milk quality.

If any one of these is below average, margins will suffer. Milk production in the future will certainly be greatly affected by the changes in the EC Common Agricultural Policy. Profits, *or even survival,* in the future will depend increasingly upon a greater appreciation and application of dairy management.

CHAPTER 17

PROFITABILITY IN MILK PRODUCTION

The main purpose of this chapter is to review the factors which determine the profitability of milk production.*

These factors can be divided into three groups. These are:

1. the economic and political climate;
2. the agricultural and physical characteristics of individual farms;
3. the differences in the systems of husbandry and management practices adopted.

ECONOMIC AND POLITICAL FACTORS

The economic and political factors of this country, and in more recent times of the other EC countries, influence and determine the price the dairy farmer receives for his milk, the price he pays for purchased feeds, and the cost of other inputs such as labour and machinery. The individual farmer has very little influence on these factors and there is relatively little he can do about them on his own. There are, however, national organisations working on his behalf such as the National Farmers Union and the Milk Marketing Board. It is to organisations such as these that he has to turn to and give his support if he wishes to seek more favourable milk price-to-cost ratios.

An examination in depth of these national and international factors is not the main purpose of this chapter. It has to be stressed, however, that the success or otherwise of government policy, and the influence or otherwise which dairy farmers have on this policy, has a very considerable effect on the profitability of milk production.

* Obviously this cannot be considered in great detail in the space allowed here and for more extensive discussion readers should refer to *Dairy Farm Business Management* by Slater and Throup, published by Farming Press Books.

Factors Determining Milk Price

Historically, milk production has been one of the most profitable enterprises in British farming. Generally it is regarded as more profitable than beef and sheep production and it accounts for a very significant proportion of the total national farm output.

Prior to entry into the EC in 1973, the price of milk and hence the profitability of dairy farming was very dependent on the proportion of milk being used in the liquid (i.e. retail) milk market, as opposed to the less profitable manufactured market, particularly butter. Manufactured milk could be, and was, imported at low prices. Consequently any substantial expansion in milk production above the liquid market requirements led to a significant fall in producers' returns. Government policy at that time was largely based on the liquid market and the concept of 'Standard Quantities' became established. If production exceeded the standard quantity, then the price received by producers fell due to the low price received for manufactured milk.

Since our entry into the EC, the import of milk products at low world market prices has not been possible due to the EC system of levies on imports. The price received for milk used for manufacture has increased and this has allowed the home production of milk to be increased without any substantial fall in prices.

These trends are illustrated by information taken from the 1989 edition of *Dairy Facts & Figures* a publication produced annually by the Milk Marketing Board.

Table 17.1 shows the usage of milk in the UK since 1954. During the period from 1954–55 to 1988–89 consumption of liquid milk rose to a maximum in 1974–75 and then dropped steadily to its present level of 5,910 million litres in 1988–89.

As the liquid market decreased so the milk manufacturing market increased to use the surplus milk.

Effect of Common Market on Milk Price

Entry into the EC Common Agricultural Policy (CAP) occurred when the UK joined the European Community in 1973.

There was a five year transitional period when the guaranteed price arrangements in the UK were replaced by the EC Market Support system. This brought about the conditions whereby the price of manufacturing milk increased substantially. Although the UK government no longer guaranteed the producer price,

Table 17.1 Utilisation of Milk Produced off Farms in England and Wales, 1954–89

Year	Liquid		Manufacturing		Total
	(million litres)	%	(million litres)	%	(million litres)
1954–55	6,103	81	1,413	19	7,516
1964–65	6,647	73	2,398	27	9,045
1974–75	6,878	62	4,273	38	11,151
1984–85	6,084	48	6,505	52	12,589
1988–89	5,910	51	5,655	49	11,565

it still retained its control over the liquid milk price, allowing some influence over producer price. These controls eventually disappeared by December 1984.

Since that date, the final milk price in each MMB area has been negotiated by a joint committee consisting of members of the MMB and representatives of the Dairy Trades Federation (an association of milk buyers). The setting of the milk price in the UK by the joint committee resembled the arrangements in operation before the War except that the basic milk price was set by the EC Common Agricultural Policy.

Common Agricultural Policy—Dairy Sector

The determination of milk price in the EC will in the future depend upon the marketing of dairy products within the community securing product prices which will enable milk producers to obtain a 'target price' for milk.

The target price is defined as a 'delivered to dairy' price for milk of 3.7 per cent butterfat content. It is not a guaranteed price, rather it is a means whereby the CAP attempts to establish market conditions which will allow the target price to be more easily realised.

Steps taken to achieve these market conditions in the community consist of:

(i) Use of import levies to prevent dairy produce entering community markets below the threshold or minimum import price.

(ii) Provision of subsidies on exports to enable community products to compete with exports from non EC countries in international markets.
(iii) Provision of a base to the market for dairy products within the Community by means of a system of intervention buying of butter and skimmed milk powder.

Ideally, the intervention price supports the milk price at 92.6 per cent of the target price.

The market is further helped by payment of subsidies on milk and certain dairy products consumed by school children.

The target price and intervention prices for both butter and skimmed milk powder are fixed each year by the EC Council of Agricultural Ministers in their annual round of price negotiations.

Intervention and Milk Quotas

Due to the favourable milk price awarded through the EC Common Agricultural Policy, the supply of milk in the EC rapidly outstripped demand, resulting in increasingly large amounts of butter and skimmed milk powder being put into intervention stores. To cut back production, quotas were introduced in 1984, whereby a supplementary levy (super levy) is charged on milk produced in excess of the quota.

The quota applies to virtually all sales of milk or milk products from farms. The dairy wholesale quota applies to deliveries to processing dairies. A direct sales quota applies to milk and milk products provided by farms direct to consumers.

In 1987 steps were taken by the EC Agricultural Ministers to limit intervention buying for butter and skimmed milk. The agreed levels of 180,00 tonnes for butter and 100,000 tonnes for skimmed milk powder were reached by mid 1987 and since then intervention has been virtually suspended.

Needless to say, as a result of the EC agricultural legislation, the ultimate control for fixing the milk price has been taken from individual countries, and transferred to the Common Agricultural Policy (CAP).

This is obviously a massive step removed from the traditional milk price discussions which took place at a comparatively 'parochial level' in the UK during the 1930s.

Farmers can be forgiven for thinking that control over their own destiny, i.e. setting their own milk price, will mainly pass out of their hands altogether in the forseeable future, and be increasingly affected by political pressures.

The continued change in the market outlet, i.e. towards a greater proportion going to manufacture, also has implications from a milk composition point of view. As the proportion going into manufacture increases so does the significance of the butter-fat and solids-not-fat content of the milk. This led to the introduction by the Milk Marketing Board in April 1984 of a new quality payments system.

The increase in the proportion of milk used for manufacturing purposes was a major reason for the introduction in 1982–83 of stiffer penalties for milk contaminated with antibiotics. Milk contaminated with even very low levels of antibiotics leads to very substantial losses in manufacture, particularly in cheese-making.

Currently much emphasis is being placed on the need for farmers to pay more attention to marketing. For many years dairy farmers have been fortunate to be able to concentrate on the job of producing milk and leave the job of marketing to the Milk Marketing Board. Hopefully this state of affairs will continue in the future but the milk producer must heed the market and price signals coming from the board. These are expected to continue to place more emphasis on the need for quality and less emphasis on winter milk production. The producer should alter his production methods accordingly.

Attention up to now has been focused on factors determining the milk price. Economics and political factors also determine the price received by a dairy farmer for his main by-products, i.e. cull cows and calves. These prices can fluctuate widely from year to year due to variations in the profitability of beef production. The significance of the income from these two sources is often under-estimated. Periods of hardship for dairy farmers tend to coincide with periods of low profitability in beef production, e.g. in 1974–75 and 1983–84. Similarly periods of prosperity tend to coincide, e.g. 1972–73, and 1982–83.

A useful rule-of-thumb is that if price for beef cattle per kilogram approaches the price of barley per tonne, then good calf prices will be obtained, e.g. if beef price is 105p/kg when barley is £115/tonne.

A recent development which will certainly affect dairy farming profits related to the beef industry was the occurrence and spread of Bovine Spongiform Encephalopathy (BSE). This is a disease of the cow nervous system.

The possibility of infected meat being sold and infecting humans, led to a drastic decline in beef eating habits in the UK and a virtual standstill in beef exports to other EC countries. This resulted in very low prices for cull cows, calves and fat cattle.

This incident demonstrates how the economics of dairying are subject to pressure from external forces.

The FMS Information Unit of the MMB has provided a total farm business costing service for many years. The annual summaries are made up of records from just over 100 farms, all of which are heavily committed to milk production. The physical details concerning the farms are found in Table 17.2.

Table 17.2 Physical Details of Farms

Year ending—spring	*1986*	*1989*
Farm size (ha)	88	96
Number of cows	114	118
Number of other cattle (LSU)	47	51
Yield (litres/cow)	5,365	5,361
Concentrate use (kg/cow)	1,557	1,433
Concentrate use (kg/litre)	0.29	0.27
Stocking rate (LSU per ha)	2.00	1.94
Nitrogen use (kg/ha)	266	265
Replacement rate %	21	21
Arable area (ha)	7	8

Source: FMS Costed Farms, MMB, 1989.

After the traumatic effect of the introduction of quota in 1984, when levels of production received a severe blow, many dairy farmers were pessimistic. However, over recent years, some farmers have increased their milk production potential by purchasing or leasing extra quota. Yield per cow on the farms surveyed in Table 17.2 has, however, tended to stay the same.

Farm size has increased, allowing cow numbers to go up, and there has also been a slight increase in young stock rearing.

Perhaps the most noticeable change shown in the table is the reduction in concentrate usage. With milk production being cut back by quotas, so there has been increasing interest in improving the margin over feed per litre. This can only be achieved by making better use of home grown forage thereby allowing concentrate usage to be reduced from 0.29 to 0.27 kg/litre and from 1,557 kg to 1,433 kg/cow. Figures from the table could be used as standards for comparison with your own farm. It is suggested, however, that farms providing the survey figures have larger herds, and higher yields than the national average.

Having established that the production details of the survey farms have continued at a fairly steady level, attention can now be given to the trends in profitability of the farms, see Table 17.3.

Table 17.3 Analysis of Farm Financial Details, 1986–1989

Enterprise Gross Margins	*Year ending spring* *1986*	*% of Total*	*1989*	*% of Total*	*Compare 86–89 % Change*
Dairy herd	57,018	85.7	78,053	84.0	+36.8
Young stock	3,536	5.3	4,360	4.6	+23.2
Other livestock	1,347	2.0	2,406	2.6	+78.6
Arable	2,695	4.1	2,800	3.0	+3.8
Other farm income	1,933	2.9	5,357	5.8	+277
Total Gross Margins	*66,529*	*100.00*	*92,976*	*100.00*	*+39.75*
Overhead Costs					
Wages	11,605	20.69	13,650	21.11	+17.6
Power & machinery:					
Cost	13,543	24.14	16,153	24.99	+19.3
Depreciation	4,849	8.65	5,725	8.85	+18.1
Sundries	6,106	10.88	7,280	11.29	+19.2
Property charges:					
Cost	7,334	13.07	9,483	14.67	+9.9
Depreciation	1,942	3.47	2,308	3.57	+8.8
Interest charges	10,714	19.10	10,038	15.52	+9.3
Total overhead costs	56,093	100.00	64,637	100.00	+15.2
Profit	10,436		28,339		+271
Management and investment income	6,380		20,524		+321
Tenant's capital	120,886		121,892		+0.08
MII return on tenant's capital	5.3		16.8		+316

Source: FMS Costed Farms, MMB. 1989.

Comparison of the years ending 1986 and 1989 indicates several trends:

1. The dairy herd and livestock enterprises improved gross margins relative to arable and other farm income. This detail indicates that the dairy sector remained the strongest area of UK agriculture. This was illustrated particularly by the number of farmers who left dairying during 1977–81 under the EC-Non Marketing of Milk Scheme only to return to dairying as a result of the SLOM scheme which made 600,000 tonnes of quota available in 1989–90.
2. The dairy herd gross margins increased by 37 per cent during 1986–89. This represented 84 per cent of the total farm gross margin.
3. During 1986–89, overhead costs were kept under reasonable control: 15.2 per cent increase over 3 years. This suggests an annual figure closely in touch with inflation. This is important since profits are a result of gross margins less overhead costs.

   It is interesting to observe the way in which the total overhead charges were allocated. Wages have been under constant pressure since quotas were brought in. However on the majority of farms slack time has been used more effectively during recent years, helping to keep the annual increase in wages in line with the annual rate of inflation. Increases in machinery costs have also been controlled and are moving towards annual rates similar to inflation.

   Property charges also follow the pattern of moving towards inflation. Charges under this heading include rent, rates, repairs to property and depreciation on fixed equipment. Interest charges rose rapidly during 1988–89 from 5 to 15 per cent.

   The amount of interest paid did not, however, rise in line with the increasing interest rate since on most farms extra income had been used to reduce various loans/overdrafts.
4. Profit during the period increased dramatically. While total farm gross margin increased by some 39 per cent, fixed costs increased by only 15 per cent.

Increased income with little change in costs is a very simple recipe for success. This poses the question concerning how to increase the dairy gross margin.

The reason for this success in the three years illustrated is largely due to the interaction between milk and feed prices. This relationship is referred to as the **milk price/concentrate price ratio**. A ratio of 1:1 means that the price of a litre of milk @ 15p/litre is

Table 17.4 Milk Price / Milk Concentrate Ratio 1986–89

	1986	*1987*	*1988*	*1989*
Milk price (pence/litre)	15.49	15.88	16.40	17.79
Concentrate (pence/kg)	13.7	12.9	12.8	13.7
Milk/concentrate price ratio	1.13	1.17	1.28	1.29

the same as 1 kg of concentrates at 15p/kg (or £150/tonne).

Table 17.4 clearly illustrates that the milk price/concentrate price ratio has been increasingly favourable for dairy farmers in England and Wales during the period 1986–89 with the effect of quotas and emphasis on milk from farm resources, so the concentrate price has remained low while the milk price has steadily increased. The influence of the milk price/concentrate price ratio on gross margin is shown in Figure 17.1.

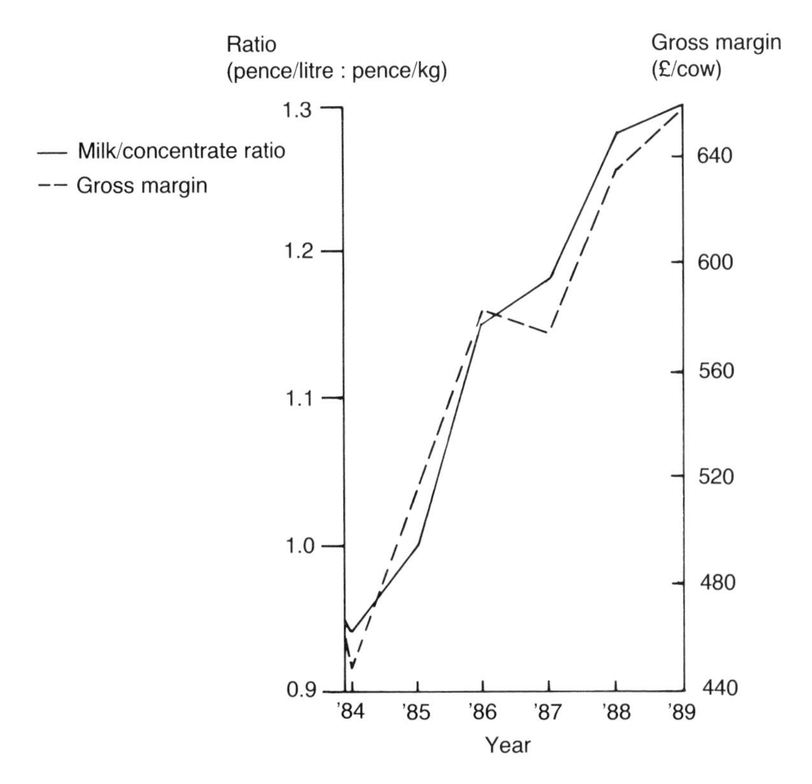

Figure 17.1 Trends in milk/concentrate ratio and gross margins
FMS Information Unit, MMB 1988–89

Another relationship which has become increasingly important is the connection between milk price and wages. The general level of wages in the country has risen during the past thirty years. This rubbed off in the dairy industry to a limited extent, but for many years was not equalled by appropriate milk price increases. This inevitably resulted in farm labour being cut back in an attempt to cut costs. A way of reducing man hours per cow was to increase the mechanisation and building investment. Unfortunately many farmers were not able to afford the cost of modernisation and gave up milk production. The continuing downward trend in the number of milk producers in England and Wales is shown in Table 17.5.

Table 17.5 Milk Producers, England and Wales

Year	*No. Producers*	*No. Dairy Cows (thousands)*	*No. Cows per Producer*	*Yield per Cow litres*
1960	123,137	2,595	21	3,320
1965	100,449	2,650	26	3,545
1970	80,265	2,714	34	3,755
1975	60,279	2,701	45	4,070
1980	43,358	2,672	62	4,715
1985	37,815	2,580	68	4,765
1989	32,395	2,341	72	4,915
1989 as a per cent of 1960	26	90	342	148

Source: Dairy Facts & Figures, MMB 1989.

The number of milk producers in England and Wales fell by 74 per cent between 1960 and 1989. Similar trends took place in Scotland and Northern Ireland.

Cow numbers remained fairly steady until 1984 when quotas came in, after which they declined. The number of cows per producer rose steadily through the whole period 1960–89 from 21 to 72, which is in line with the comments on labour use and increasing mechanisation.

The overall conclusion must be one of slight reductions in cow numbers accompanied by much larger reductions in the number of dairy farms, i.e. milk producers.

Table 17.6 Distribution of Cows by Herd Size (England and Wales)

Size group *No. of Dairy Cows*	*1982* *% of Cows*	*1988* *% of Cows*
6–49	22.1	19.6
50–99	39.4	39.3
100–199	28.1	30.8
200 and over	10.4	10.3
Total	100.0	100.0

Source: Dairy Facts & Figures, MMB 1989.

The number of cows needed to make a dairy farm viable is likely to continue to increase, probably resulting in the disappearance of more small farms.

The expected increase in large dairy farms has not materialised in recent years. The cost of labour is very high on large farms and even more importantly, there is an increasing shortage of skilled dairy staff.

Bearing these factors in mind it would seem that the family farm with a 100–150 cow herd is likely to be best equipped to face the pressures of the next decade.

AGRICULTURAL AND PHYSICAL CHARACTERISTICS OF THE INDIVIDUAL FARM

Reference has already been made to the substantial increase in the size of the average herd. Many small producers have been forced to give up milk production and this trend is expected to continue. The very small farmer often has inherent disadvantages in terms of buildings and other fixed equipment. Consequently he is uncompetitive from a labour productivity point of view.

The availability or otherwise of suitable buildings and fixed equipment is, however, a major factor affecting the viability of a dairy enterprise irrespective of the size of the farm.

In this context it needs to be noted that the rent payable for a farm at any particular point in time or its equivalent (i.e. interest charges on borrowed capital) often does not reflect the adequacy or otherwise of the fixed equipment and buildings. The 'rent equivalent' reflects many factors but the most important is often the length of time which the farm has been rented or owned; the longer this period the lower the rent equivalent tends to be. Consequently a farm with a high rent equivalent may have poor buildings and fixed equipment whereas a farm with a relatively low rent equivalent may have good fixed equipment and buildings. Young farmers in particular are faced with this problem. To overcome it they have to be prepared to accept a lower level of personal wages from their farm than they would receive as hired workers doing the same work.

The lack of adequate buildings and fixed equipment tends to result in higher than average labour costs. This can be overcome by making the appropriate investments, but in the short term this leads to a substantial financial burden being placed on the business. This makes the business less competitive compared to established neighbouring businesses even though it may be just as well or better managed.

The lack of good buildings, good milking premises and good services such as water and electricity is considered to have been one of the main reasons for producers giving up milk production during the past twenty-five years. It is also considered that this factor will continue to be of vital importance in the future.

The quality and nature of the land is also a significant factor in milk production, but the significance of this factor is sometimes overestimated. The milk producer at high altitude in the north with a short grass-growing season obviously has a handicap compared to a lowland farmer in the southwest. Many costings, however, have shown that differences in location are not the most significant factors determining the difference in profit between individual farms. Profits achieved in one part of the country do not differ significantly from those in another. There is a trend, however, for dairy farmers to become more concentrated in the west of the country compared to the east.

HUSBANDRY AND MANAGEMENT FACTORS

A feature of all economic surveys of milk production is the very wide variation in profits between farms. These differences are found even when the surveys are restricted to small geographical areas and to farms carrying out similar systems of husbandry, e.g. all producing milk from Friesian cows on a silage-based feeding system.

Considerable discussion takes place as to which is the best system of producing milk, but the first question a dairy farmer or herdsman should ask is, 'Am I getting the best I can out of my present system?' He should follow up with, 'If not, what can I do about it?' rather than looking to a new system to solve his problems.

In this section it is necessary to quote costs and returns to highlight the significance of various factors. This immediately causes problems:

- Financial results differ considerably from year to year as has already been mentioned so the figures shown may not be normal.
- We are living in a period of rapid inflation and quoted costs are out of date within a very short time, in fact they are often out of date by the time they are published.

  Financial data quoted in this book is almost certain to be out of date by the time it reaches the reader. It should be used therefore not as a guide to actual levels of profit but as an indicator of the main factors that determine the profit level.

POINTERS TO PROFIT

Results per cow are given in Table 17.7 and can be used to see how a typical farmer's profit is composed. Gross output is made up of:

1. Milk Sales

This accounts for some 92.4 per cent of gross output.

Value of milk per cow = number of litres × milk price/litre.

The number of litres is limited by quota but on any farm where milk is being produced economically then additional milk production potential can be obtained through leasing or buying in extra quota. Outlay of money for this purpose has to be looked at in relation to the current cash flow situation and any overdraft commitments. Milk price per litre is increased by good milk quality, requiring correct levels of feed to be given throughout the lactation. Any review of efficiency factors leads to a review of the quality of the associated husbandry. Yields and milk quality can only be controlled by the farmer. Solutions depend upon better husbandry.

Table 17.7 Results per Cow

Year ending spring	*1988*	%	*1989*	%
Milk sales	880	94.1	953	92.4
plus Calf output	97	5.8	114	7.6
less Herd depreciation	42		35	
Gross output	935	100.0	1,032	100.0
Variable costs				
Concentrates	189	54.7	196	52.4
Purchased bulk	11	3.2	10	2.7
Forage	78	22.5	83	22.2
Sundries	68	19.6	85	22.7
Total variable costs	346	100.0	374	100.0
Gross margin per cow	589		658	
Margin over concentrates (MOC)	691		757	
Overhead costs				
Wages	109		115	
Power and Machinery	181		185	
Other costs	230		246	
Total Overhead costs	520		546	
Profit	69		112	
Yield	5,367		5,361	
Milk price	16.40		17.79	

Source: FMS Report, MMB, 1989.

2. Calf Output

The emphasis on how to achieve a high calf output has changed through recent years. Calf sales contribute significantly to dairy profits. A first step in recent years has been the introduction of new types of crossing bulls. Continental breds, e.g. Charolais, Limousin, Simmental, Belgian Blue, have all been and continue to be very fashionable. It has been quite usual for cross-bred continental calves out of Friesians to make a premium of £50–£100 at local markets. The majority of such calves are sold within a few days of birth.

Calf mortality has been under control for many years with veterinary help. Lack of calf sales can in some cases be attributed to a poor herd fertility and high calving interval. The aim should be one calf per cow per year.

3. Herd Depreciation

The first requirement for a low depreciation rate is for the cows to have long productive lives. If the cull rate is brought down then automatically fewer heifer calves will be needed and surplus heifers will be available for sale.

Once this position is reached then two alternatives are available:

(i) maintain a low cull rate—introduce relatively few heifers to the herd and sell the surplus;
(ii) deliberately increase the cull rate—if more older, lower producing, cows are culled, then the herd production potential can be increased by bringing in more heifers from better proven blood lines.

Herd depreciation, as a cost item, depends upon the cull price received for cows and the market price for bought in heifers—the going market price should ideally also be applied to home reared heifers.

For example on a 100 cow herd.

	£
Average value of 25 heifers or cows purchased 25 @ £700 each =	17,500
Average price of heifers or cows sold 25 @ £450 each	11,250
Herd depreciation	6,250

Depreciation per cow in the herd = $\frac{6,250}{100}$ = £62.50/cow

Herd depreciation has been low in the late eighties due to a very firm market for cull cows provided by continental buyers which has resulted in prices occasionally topping £600 per cow.

With the appearance of BSE in the UK, the export market has dropped away quite severely, thus lowering cull cow prices.

Herd depreciation is affected by seasonal weather fluctuations.

In a run of dry summers, milk production is down on quota. Farmers therefore look for extra milk cows/heifers to fill their quota, forcing up prices for newly calved heifers and cows.

These outside factors are not within the control of the farmer. Nonetheless timeliness of selling and buying can help to offset adverse market levels.

A recent introduction on some farms has been the feeding of cull cows on silage for a month before sale in order to improve body condition, thereby leading to a better cull price.

Variable costs

These include:

Concentrates These cover two categories:

(i) Purchased concentrates from commercial feedstuff firms.
(ii) Home grown concentrates e.g. barley, wheat, peas, beans, etc.

Concentrates make up 52.4 per cent of the variable costs of dairy production. Any economies which can be operated in this area will have a considerable effect on year-end profits. The main purpose of concentrate usage is to reach the target yield by using as little or as much concentrate as is required.

Mention has previously been made concerning the need to achieve the best balance between concentrates and bulk feed in order to obtain the best ongoing profit.

The correct balance is very variable, depending greatly on the available quantities and qualities of bulk food grown on the farm. Climatic effects can bring about variations of between £50 and £100/cow on the margin over feed at the year-end. With higher levels of milk yield it is sometimes desirable to make use of a feed fence in addition to in-parlour feeding. Often straight concentrates are used on a feed fence, e.g. cereals or sugar beet and some form of protein concentrate—maize gluten meal, dark distiller's grains or soya bean meal.

2 kg of sugar beet	@ 11p/kg =	22
2 kg of maize gluten	@ 12p/kg =	24
		46

giving an average price of 11.5p/kg compared to a proprietary concentrate at 15p/kg. There are many alternatives for cheapening a ration—the essential end result is the maximum possible economy while still obtaining the yield level required. Obviously though, care must be taken to maintain mineral and vitamin levels.

Purchased Bulk Feeds Normally plans are made to produce all necessary bulk feed from the farm. Occasions which make it necessary to purchase bulk feed occur:

1. when there are extreme weather conditions restricting grass growth, making it necessary to top up farm produced bulk with purchased brewer's grains, pressed pulp, potatoes, etc.;
2. when silage is made in very wet conditions, becomes butyric and is therefore quite unpalatable. In such circumstances, purchased bulk feed, e.g. brewer's grains, would dilute the silage, making it more palatable and therefore assisting intake.

   Dairy farms in the dry eastern parts of the country often have a policy of regularly using crop by-products (e.g. vegetable waste, surplus carrots or cabbages) which may be available at give-away prices, allowing the more expensive homegrown silage to be left in store until a later date.

Forage Costs This item covers the cost of establishing grassland, e.g. seeds, fertilisers and sprays plus the cost of the regular fertiliser applications. This would cover the spreading of N, P and K. Levels of N application can vary considerably from farm to farm; soil type and climate accounting for greatly differing requirements.

Research has suggested that regular planning of fertiliser application is a good practise. In earlier chapters there was considerable discussion on stocking rates, and their association with the fertiliser. Intensive stocking rates, e.g. 2.2–2.5 cows/kg require the application of high levels of N.

With the decline in the profitability of arable crops grown on dairy farms, there has been a move towards putting a larger area down to grass, lowering the stocking rate and thus allowing the level of fertiliser applications to be reduced. The current spate of dry summers has also led to the review of the need to apply fertiliser.

N fertiliser certainly has a very effective role to play in growing large grass crops in the spring for silage making. Recent experience has shown that heavy N applications (i.e. 110–120 units) can also be responsible for high ammonia content in the silage. As always, it is advisable to show caution with the amount of nitrogen applied—excessive application levels can cause problems.

Grass, whether fresh or for conservation, should be grown to satisfy the farm livestock requirement with an appropriate reserve for late springs, wet autumns, dry summers, etc.

The idea of 'buffer feeding' i.e. feeding silage at grass is very sound. It is the best way of finding out whether the grass supply is adequate for the cows requirements.

Systems change over the years. At one time 7–8 tonnes of silage per cow per year was considered adequate. Now, however, it is realised that 10–12 tonnes may be needed to ensure adequate availability of silage in both winter and summer.

Economies in forage costs are possible but must not detract from the purpose of providing adequate bulk feed throughout the year.

Phosphorus and potash may also need to be applied. Shortages can be identified by regular soil analyses.

Fertilisers are manufactured to specific chemical recommendations. It has never been claimed that any basic fertiliser had the 'X'—unknown factor—as with feeding stuffs. Therefore any economies are only going to be possible by paying attention to the unit price of the nutrients. Physical condition of the fertilisers can be variable.

Sundries It is perhaps surprising to find that sundries account for some 22.7 per cent of variable costs.

Items included under this heading are veterinary and medicine bills, dairy stores, AI, milk recording, herd registration fees, and bedding, plus any other small items difficult to allocate. Some economies are possible to improve profit.

Healthy herds require less veterinary work. Good husbandry practices therefore help reduce vet's bills. Although drugs tend to be expensive, when used wisely they are highly valued. Occasionally they could be used to make up for poor management.

Awareness of the potentially high costs involved should concentrate attention on keeping the levels down.

Margin over Concentrates (MOC)

This is the most widely accepted first measurement which can be used as a check on overall enterprise efficiency. It is defined as:

MOC = milk sales − concentrate cost.

It is calculated using the two basic sets of information concerned with:

1. milk sales: how much milk at what quality was produced.
2. concentrate cost: how much concentrate was fed to achieve the result.

Farmers are not naturally inclined to recording facts on the farm. In this instance, however, the farmer has little to do except to keep both his monthly milk statement and monthly feed bills.

Table 17.7 shows that the MOC was £691 per cow in 1988, rising to £757 in 1989. These figures should be looked at in the context that the MOC for cows generally lies within a range of £600–1,200 per cow. The MOC can therefore differ by up to £600 per cow. With a 100 cow herd this would give farmer 'A' £60,000 more profit than farmer 'B'.

Obtaining the best possible total herd margin depends on detailed consideration of each of the factors summarised below:

Milk sales affected by

Yield:
- Nutrition
- Breed
- Seasonality of calving
- Fertility
- Health
- Housing

Price:
- Quality
  - —Butter fat
  - —Protein
  - —Lactose
- Nutrition
- Hygiene
  - —TBC
- Stage of lactation
- Age

Concentrate cost affected by

- Yield required
- Feeding system
- Bulk feed supply
  - —Summer grazing
  - —Winter bulk
- Quality of bulk feed throughout the year
- Stocking rate
- Concentrate type
  - —Balanced compounds
  - —Straights

Many of these factors are 'long term' and their improvement should be on-going. Breeding is a lifetime's work.

Other factors occur in certain seasons and not in others. A dry summer creates problems in both quality of grass—fibre rather than leaf—and quantity of food available.

The best MOC figure is only obtained by constant attention to detail.

Margin over Total Feed (MOTF)

This is the management way to account for farms which have limited ability, through climatic and geographic reasons, to supply farm-grown bulk feed for the herd.

Cows need to have their appetites satisfied. This has a cost which must be recorded for use in this efficiency measure.

The cost, whether for purchased brewer's grains, hay, pressed sugar beet pulp, carrots, fodder beet or baled silage, can easily be found from the invoice and can therefore be included in the total cost of feeding the dairy cow. It is not necessary, however, to know the physical amounts used, except for reference.

A word of warning concerning the application of MOC and MOPF (margin over purchased feed—defined as milk sales minus total purchased feeds). Their primary use is in measuring progress on a farm over a number of years, rather than for inter-farm comparisons; every farm is different and thus too much importance should not be placed on such comparisons.

The other variable costs involved before arriving at the gross margin are rather more difficult to allocate than feed. Fertiliser is applied over a whole farm, making it difficult to allocate exactly which enterprise the fertiliser is helping. The grassland area is often utilised in quick succession by the dairy herd, young stock and beef cattle, followed by cutting for conservation products. The silage and hay produced may be used by any of the variety of stock on the farm.

All variable costs must be under constant review but having reduced them to the most efficient level then the only opening for higher profits is increased production.

In seeking to achieve a high MOC a farmer has basically to choose between three solutions. These are (1) to increase the yield without any increases in feed costs, (2) to reduce feed costs without corresponding reductions in yield, or (3) to increase yields and reduce feed costs at the same time. The latter may seem impossible but in practice some high-yielding herds do achieve their yields with lower than average concentrate feed costs. This is due to their ability to combine above-average stockmanship, resulting in good yields, with above-average grassmanship resulting in good-quality grazing with bulk foods. The most successful farmers are those with the ability to combine both facets in their management strategy.

Improving the milk price is an alternative method of trying to improve the margin. As mentioned previously this aspect may become of increasing significance in the future but to date this factor has been of secondary importance compared to the need to attain high yields.

When assessing the MOC obtained on a particular farm it is necessary to also take into account the stocking rate, together with the level of expenditure on forage costs, and fixed costs associated with the production of the forage. This is an exercise requiring considerable knowledge and experience in farm business management and costings, and there is not sufficient space to allow all these aspects to be covered in this chapter.

It is necessary, however, to emphasise the significance of MOC as an efficiency measure in dairy farming. In turn this leads to a need to underline the main factors resulting in a good yield and a good margin. Foremost among these is the quality of grazing and bulk feed available on the farm. Not only does this keep down feed costs, it also helps the attainment of good yields.

The various factors determining milk yield and efficiency of grassland and forage utilisation have been detailed elsewhere in this book. The attainment of good margins depends largely on how well these factors are controlled and implemented on the farm. Mention has been made elsewhere of various recording systems and again this is considered a vital factor determining the level of margin attained. All dairy farmers should monitor carefully their yields and feed inputs in both physical and economic terms.

Return per Hectare

Emphasis up to now has been placed on margins per cow rather than per hectare. This can be justified since the capital invested in a dairy farm tends to be closely related to the number of cows. Consequently high margins per cow tend to be associated with high returns to capital.

A large proportion of dairy herds, however, are on small farms, and on small farms land is the most limiting factor and consequently return per hectare becomes more important than return per cow. The stocking rate in this context becomes a most important factor. The most successful dairy farmers achieve not only higher margins per cow but manage to achieve this at higher stocking rates. This is illustrated in Table 17.8. Note that the expenditure on concentrates per cow and forage per hectare has been purposely set at the same level for both the above average and the average farmer. The secret of the above average farmer's success is that he achieves a greater output for the same level of inputs. This applies whether the inputs are feed or fertilisers, and is a reflection of his increased managerial skill and ability to do things at the right time. Silage quality, for example, is not determined by the amount of fertiliser used but on the ability to cut and conserve grass at the right stage of growth together with the appropriate degree of wilting and consolidation.

Table 17.8 Stocking Rates and Margins per Cow

	Average		*Above Average*	
Results per cow		£		£
Milk sales	5,300 l @ 17p/l	901	6,300 @ 17p/l	1,071
Concentrate feed costs	5,300 l @ 3.69p/l	196	6,300 @ 3.11p/l	196
MOC		705		875
	Margin/litre 13.3p		Margin/litre 13.9p	
Forage costs				
£160/ha	0.5 ha	80	0.45 ha	72
Margin over feed and forage per cow		625		803
Results per hectare				
No. of cows per ha		2.0		2.2
Margin over concentrate and forage costs per ha (MOTFF)		1,250		1,766

This ability to do things at the right time and by the correct methods has a very marked effect on the profitability of dairy farming. Not only do the best farmers achieve higher yields and stocking rates from given feed and fertiliser inputs, but they also achieve these with no greater expenditure on labour and machinery. Consequently these are the farmers who have the ability to pay high rent or service high rent-equivalents in the form of finance charges.

Table 17.9 Stocking Rates and Profits

	Average *£ per cow*	*Above Average* *£ per cow*
Milk sales	901	1,071
plus calves	+90	+90
less depreciation	−60	−60
Gross output	931	1,101
Less		
Feed	196	196
Forage	80	72
Vet and sundries	60	60
Total variable costs	336	328
Gross margin	595	773
	per hectare	*per hectare*
No. of cows/ha	2.0	2.2
Gross margin/ha	1,190	1,700
Fixed costs		
Labour	450	450
Power and Mach.	300	300
Sundry overheads	270	270
Total fixed costs	1,020	1,020
Profit before rent and finance	170	680

This is illustrated in Table 17.9. By making better use of resources an above-average farmer doubles the profit before rent and finance charges. This allows him to service a well-above-average rent charge and still make an above-average management and investment income.

In Table 17.9 it will be noticed that no difference has been assumed between the average and above-average farm in terms of calf values and herd maintenance costs.

As mentioned previously, calf values have a very significant effect on the fortunes of dairy farming from year to year, but they account for very little difference in profit between farms in a given year. Nonetheless it is essential to avoid calf mortality and to try to get as good a market value for calves as is possible.

Note.
The above points are made in relation to the national herd which is largely dominated by the Friesian and Canadian Holstein breeds. Significant differences do occur when comparing Friesian herds to Channel Island herds but that is not the point being made in this section. This lack of good calf output is a point that has to be carefully considered when making the decision whether or not to go for milk yield or quality.

Table 17.8 shows the considerable difference between an average or low performance herd with a low intensity stocking rate (MOTF = £1,250 per hectare) and a herd of above average performance linked with a higher stocking rate (MOC = £1,766 per hectare).

When fixed costs are included in the equation then the farm profit again is greatly improved with the more intensive stocking rate. In spite of this the total farm profit shown in Table 17.9 show a very poor return. This is a fact that has been evident for many years, when the profits are examined from a commercial viewpoint. Farmers are very good husbandmen, craftsmen and husbandry experts, but very few merit the title businessman, i.e. a person who does things to make a 'profit'.

Farming is still very much a way of life. Surveys by the MMB over many years have shown a very poor return on tenant's capital compared to the going bank rate. This is well illustrated in Table 17.10.

Table 17.10 Comparison Between Return on Tenant's Capital and Bank Interest Charges, 1981–89

	81	*82*	*83*	*84*	*85*	*86*	*87*	*88*	*89*
Return on Tenant's Capital	6.1	7.7	8.8	nil	3.6	5.3	5.2	11.4	16.8
Annual Bank Base Rate (Interest Charge)	13.24	11.93	9.81	9.75	12.45	10.89	9.15	10.16	13.8
+ Service Charge @ 2.5%	15.74	14.43	12.31	12.25	14.95	13.39	11.65	12.66	16.3

Banks impose a rate of 1½–3 per cent interest above Base rate for bank services. The amount over the base rate is fixed with each client with reference to size and fluctuations in his overdraft. When profits fall, farmers tend to worry and begin to look for other ways of managing their dairy herds to make profits and improve their financial position. Once profits begin to increase then the urgency to improve tends to fall away.

In 1989–90 a sample of good dairy farms achieved a return of only 16.8 per cent, when the bank base rate was 15 per cent + 2½ per cent = 17.5 per cent. In this situation there is no opportunity for any farmer to sit back and say 'We're doing alright.'

To make a return that barely equals the current interest charge on overdraft is not a good business habit. It is likely that in many other industries, investors would sell the business and get out. The reason that farmers don't do this is because they like farming—but hobbies don't usually pay and, in fact, are often expensive past-times.

Alongside the massive requirement of technical knowledge required to be a good farmer, must be developed a business instinct to always achieve an improving financial return on the business.

APPENDICES

APPENDIX I

Breed Comparison with Reference to Milk Yield, Butterfat and Protein Levels

	Milk Yield		*Butterfat*		*Protein*	
	1987–8	*1988–9*	*1987–8*	*1988–9*	*1987–8*	*1988–9*
Recording year	*kilograms*		*per cent by weight*			
Ayrshire	5,184	5,260	3.96	3.92	3.32	3.29
British Friesian	5,751	5,800	3.89	3.86	3.22	3.20
British Holstein	6,443	6,483	3.86	3.82	3.16	3.14
Brown Swiss	5,779	5,799	3.77	3.66	3.40	3.40
Dairy Shorthorn	5,006	5,175	3.71	3.67	3.27	3.26
Dexter	2,347	2,308	4.17	4.08	3.38	3.45
Guernsey	4,224	4,275	4.68	4.66	3.56	3.54
Jersey	3,978	4,047	5.34	5.31	3.82	3.80
Red Poll/British Dane	3,994	4,228	3.65	3.78	3.32	3.29
Simmental	3,999	4,306	4.00	3.87	3.31	3.32
South Devon	3,522	3,575	4.01	4.01	3.61	3.52
Mixed and Others	5,724	5,766	3.88	3.84	3.21	3.19
All Breeds	**5,707**	**5,758**	**3.91**	**3.88**	**3.23**	**3.21**

Results from Recorded Herds, England and Wales. The recording year is 12 months ending September. (UK Dairy Facts and Figures, 1990.)

APPENDIX II

Calculations for Estimating Content Weights

Stacks of hay and straw

First, work out the cubic capacity of the stack.

For rectangular stacks multiply length by breadth (in metres) and multiply that figure by height to eaves. Then add cubic capacity of roof—obtained by multiplying the length by breadth again and multiplying that figure by half the height of the roof, measured perpendicularly from eaves to ridge.

For round stacks, square the circumference, multiply by 0.08 and then multiply by a figure equal to the height to the eaves plus one-third of the height of the roof, measured perpendicularly from the eaves to the peak. All measurements should be made in metres.

When the total cubic capacity is known, tonnage can be worked out from the following tables:

Average number of cubic metres per tonne of hay

Condition of stack	*Square stacks*	*Round stacks*
	m^3 per tonne	
Not well settled	9.0	9.8
Fairly well settled	7.5	8.3
Very well settled	6.0	6.8

Average number of cubic metres per tonne of straw

	m^3 per tonne
Wheat straw	13.5–15.0
Oat straw	15.0–17.3
Barley straw	15.0–17.3

Roots

The average weight per cubic metre of roots in clamp is: fodder-beet 545 kg, swedes 545 kg, mangolds 561 kg.

Pit silos

Work out the cubic capacity in metres—multiply length by breadth by height, taking average measurements—then reckon that each cubic metre weighs 0.664 tonne.

Tower silos

The weight of silage per cubic metre is variable, depending on the depth of silage, diameter of silo, fineness of cutting, packing, etc., but on an average 1 cubic metre equals 664 kg.

Height of silo m	*Average weight of silage kg/m³*	*Capacity of silo in tonnes when diameter of silo in metres*			
		3	4	5	6
7.92	596	37	—	—	—
8.53	615	41	74	—	—
9.14	635	46	81	—	—
9.75	652	50	90	—	—
10.36	670	55	97	153	—
10.97	686	—	106	166	238
11.58	702	—	—	178	258
12.19	718	—	—	192	277
12.80	732	—	—	206	297
13.41	749	—	—	221	317
14.02	758	—	—	234	337
14.63	769	—	—	247	356
15.24	782	—	—	262	377

APPENDIX III

Possible Analyses and Feeding Value of Compound Feeds

	Analysis per cent by weight			*Approximate nutritive value*	
Feeding stuff	*Oil*	*Crude protein*	*Fibre no more than*	*g/kg DM DCP*	*MJ/kg DM ME*
	3–5	14–18	8	130–175	12.4–12.9
Dairy concentrates	4–5.5	14–18	7	130–175	13.2–13.8
	6–8	15–21	5	140–200	14–14.4
High protein balancer	3–6	34–38	8	300–330	11.5–12.5

APPENDIX IV

Food Values*

I. Home-grown concentrates

	DM g/kg	ME DM MJ/kg	DCP DM g/kg
Beans	860	12.8	209–248
Peas	860	13.4	225
Linseed	900	19.3	208
Dried grass			
29% crude protein or over	900	10.8	113[1]
16%–19% crude protein	900	10.6	136[2]
12%–15% crude protein	900	9.7	97[3]
Wheat	860	14.0	105
Oats	860	11.5	84
Barley	860	13.7	82
Sugar-beet pulp (dry)	900	12.7	59

[1] v. leafy [2] leafy [3] in flower

II. Purchased concentrates

	DM g/kg	ME DM MJ/kg	DCP DM g/kg
White Fishmeal	900	11.1	631
Decorticated Groundnut Cake	900	12.9	449
Dried Yeast	900	11.7	381
Soya Bean Cake	900	13.3	454
Decorticated Cottonseed Cake	900	12.3	393
High Protein Cake	—	—	—
Undecorticated Groundnut Cake	900	11.4	310
Linseed Cake	900	13.4	286
Grain Balancer Cake	—	—	—
Maize Gluten Feed	900	13.5	223
Dried Distillers' Grains	900	12.1	214
Sunflower Seed Cake	900	13.3	372
Undecorticated Cottonseed Cake	900	8.7	203
Palm Kernel Cake	900	12.8	196
Coconut Cake	900	13.0	184
Malt Culms	900	11.2	222
Dried Brewer's Grains	900	10.3	145

* Now expressed as Metabolisable Energy (MJ/kg DM), Digestible Crude Protein (g/kg DM and Dry Matter (g/kg). See MAFF Bulletin No. 33.

APPENDIX IV (continued)

Purchased concentrates *continued*	DM g/kg	ME DM MJ/kg	DCP DM g/kg
Typical Dairy Cake	—	—	—
Weatings	880	11.9	129
Bran	880	10.1	126
Maize Germ Meal	900	13.2	90
Maize Meal or Flaked	900	15.6	99
Locust Beans	860	13.8	47
Molasses	750	12.9	16
Tapioca Meal	900	15.0	13

III. Succulents	DM g/kg	ME DM MJ/kg	DCP DM g/kg
Cabbage	110	10.4	100
Kale	140	11.0	123
Rape	140	9.5	144
Beet tops	160	9.9	88
Mangolds	120	12.4	58
Swedes	120	12.8	91
Grass silage	220	7.6–10.2	98–116
Maize silage	210	10.8	70
Pasture grass (4″ stage)	200	12.2	225
Pasture grass (8″ stage)	200	11.2	130
Pasture grass (mature)	200	10.0	124
Fodder beet (Pajbjerg Rex X)	—	—	—
Wet grains	220	10.0	149
Wet Beet Pulp	180	12.7	66

IV. Roughages	DM g/kg	ME DM MJ/kg	DCP DM g/kg
Hay, very good, leafy	850	10.1	90
Seeds hay, medium	850	8.9	103
Meadow hay, poor	850	7.5	45
Lucerne hay	850	8.2	166
Straws:			
Oat	860	6.7	11
Barley	860	7.3	9
Wheat cavings	860	5.9	13
Bean and pea haulm	860	7.4	26

APPENDIX V

Table for Use in Calculating Winter Rations

Quantity available	*Amounts available per head over—*			
	3 months (91 days)	*4 months (121 days)*	*5 months (152 days)*	*6 months (182 days)*
(kg)	*(kg)*	*(kg)*	*(kg)*	*(kg)*
250	2.7	2.1	1.6	1.4
300	3.3	2.5	2.0	1.6
350	3.8	2.9	2.3	1.9
400	4.4	3.3	2.6	2.2
450	4.9	3.7	3.0	2.5
500	5.5	4.1	3.3	2.7
600	6.6	5.0	3.9	3.3
700	7.6	5.8	4.6	3.8
800	8.8	6.6	5.3	4.4
900	9.9	7.4	5.9	4.9
1,000	11.0	8.3	6.6	5.5
1,500	16.5	12.4	9.9	8.2
2,000	22.0	16.5	13.2	11.0
2,500	27.5	20.7	16.4	13.7
3,000	33.0	24.8	19.7	16.5
4,000	44.0	33.1	26.3	22.0
5,000	54.9	41.3	32.9	27.5

INDEX

Numbers in italics refer to illustrations

Farming Press Books

Listed below are a number of the agricultural and veterinary books published by Farming Press. For more information or a free illustrated book list please contact:

Farming Press Books and Videos
Wharfedale Road, Ipswich IP1 4LG, United Kingdom
Telephone (01473) 241122 Fax (01473) 240501

The Herdsman's Book • MALCOLM STANSFIELD

The techniques and skills of cattle husbandry.

A Veterinary Book for Dairy Farmers • ROGER BLOWEY

Deals with the full range of cattle and calf ailments, with the emphasis on preventative medicine.

Cattle Ailments – Recognition and Treatment • EDDIE STRAITON

An ideal quick reference with 300 photographs and a concise, action-oriented text.

Calving the Cow and Care of the Calf • EDDIE STRAITON

A highly illustrated manual offering practical, commonsense guidance.

Cattle Feeding • JOHN OWEN

A detailed account of the principles and practices of cattle feeding, including optimal diet formulation.

Calf Rearing • THICKETT, MITCHELL, HALLOWS

Covers the housed rearing of calves to twelve weeks, reflecting modern experience in a wide variety of situations.

Farming Press Books is part of Miller Freeman Professional Ltd., which publishes a range of farming magazines: *Arable Farming, Dairy Farmer, Farming News, Pig Farming, What's New in Farming*. For a specimen copy of any of these please contact the address above.